Catalysts and Surfaces
Characterization Techniques

B. Viswanathan

S. Kannan

R.C. Deka

Alpha Science International Ltd.

Oxford, U.K.

Catalysts and Surfaces
Characterization Techniques
236 pgs. | 164 figs. | 40 tbls.

B. Viswanathan
National Centre for Catalysis Research
Department of Chemistry
Indian Institute of Technology Madras
Chennai, India

S. Kannan
Discipline of Inorganic Materials and Catalysis
Central Salt and Marine Chemicals Research Institute
GB Marg, Bhavnagar, India

R.C. Deka
Department of Chemical Sciences
Tezpur University
Napaam, Tezpur
Sonitpur, Assam, India

ALPHA SCIENCE INTERNATIONAL LTD.
7200 The Quorum, Oxford Business Park North
Garsington Road, Oxford OX4 2JZ, U.K.

www.alphasci.com

ISBN 978-1-84265-315-9

Printed in India

Preface

Chemistry in two dimensions is a fascinating branch of science. Surfaces are the seats of chemical reactivity in almost all heterogeneously catalyzed reactions including those that are shape and size controlled. The possibility of examining the surfaces at atomic level of resolution has enabled the identification of active sites for catalytic reactions. This has facilitated the precise design and fabrication of catalytic systems. Thus the original perception that catalyst design is most often a chance has turned out to be a rational reasoning exercise. This has become possible by the development of a number of surface analytical techniques which can probe every aspect of the surfaces like the structure of the two dimensional lattice, the chemical identity of the surface species and the bonding scheme of the adsorbed species. In fact, one of the recent Nobel Prize was awarded to Prof Ertl for his work with surface analytical techniques. Today researchers in the field of catalysis has to empower themselves the capacity to judiciously select and exploit a host of surface analytical techniques for elucidating the finest details possible of the catalyst surface.

The Department of Science and Technology (DST), Government of India has been consciously promoting the development of human resources in this area by sponsoring "Orientation Programme for Research Scholars" consistently for the past ten years. The National Centre for Catalysis Research (NCCR) at the Indian Institute of Technology Madras has been the proud promoter of this programme. The book **'Catalysts and Surfaces – Characterization Techniques'** is one of the fruits of this endeavour. The support received from Department of Science and Technology (DST) for publishing this book as well as for the orientation programme is gratefully acknowledged.

The chapters of this book have been designed keeping in mind the techniques that have been extensively exploited today in the elucidation of catalyst surfaces. Thus, in addition to the treatment of electron spectroscopic techniques, chapters on microscopic techniques and theoretical examination of the surfaces have been included. The authors earnestly believe that the contents of this book will be useful for the students and researchers in this field in spite of the limitations on the choice of the chapters and coverage in each of the chapters. The authors record their grateful thanks to their authorities, fellow teachers, coworkers and research scholars in their respective institutions who have contributed in no small measure for the successful compilation of this book.

The authors realize the limitations and short comings in making a compilation of this kind. Therefore, any suggestions for the improvement of this endeavour will be gratefully received.

B. Viswanathan
S. Kannan
R.C. Deka

Catalysts and Surfaces

Characterization Techniques

Contents

CHAPTER 1

Introduction

The available experimental facilities enable one to probe the behaviour of individual molecules under various environments. This is especially true in the elucidation of the surface species, surface intermediates or even the dynamics of surface transformations. A variety of experimental techniques are available for each of these investigations and at times one is at a loss to know what technique to use and what information to look for. There is a number of textbooks available dealing with the fundamentals of each of these techniques and it is not our intention to repeat them. The primary purpose of this monograph is to highlight the specific features of the techniques with respect to elucidation of catalytic materials, catalytic surface, and catalytic surface transformations and also lead one to understand and appreciate what technique to use when and what information to look for. There is a number of important and fundamental questions one has to clarify before one enters into a research programme in the field of surface science and catalysis and unfortunately no straight forward answers to these questions are available.

For example, one has to define what is meant by a surface, the top monolayer of the material or top few layers of the material which is at the interface between the bulk material and the surrounding medium. This brings us to the concept of interface.

❖ How do interfaces behave?

❖ Do they behave as an algebraic sum of the behaviour of the two phases?

❖ Do the phases at the interface retain their identity?

❖ If the phases are changed in configurations and structure, what is the driving force for such changes?

❖ To how many layers in each of these phases, these configurational changes are felt?

❖ From what depth or number of layers deep down from the surface or interface the bulk properties of these phases are manifested?

❖ If the surfaces and interfaces are a dynamic one, why do we need the study of the surfaces in static mode?

❖ Is it for the qualitative and quantitative elemental composition?

❖ Is it to assess whether there is any accumulation or depletion of species from other phases?
❖ Is there any accumulation of the species from one phase thus leading to binding at the surface?
❖ What is the nature of this adsorption?
❖ What is the adsorption strength?
❖ What is the structure of the adsorbed state as compared to the free molecules in its own state?
❖ How do the properties of these molecules in the adsorbed state differ from that they exhibit in their free state?

One can easily see that not only the static properties but also the dynamics of these species present at the interface is altered in such a way that one has to reformulate his expressions for transformation kinetics. This means that we are entering an area of chemistry which is different from that of the bulk properties or that of individual molecules, both of which are familiar and are easily amenable for study by chemists using a number of experimental tools at the disposal.

There is a variety of interfaces, like solid/solid, solid/liquid, liquid/liquid, solid/gas, liquid/gas and possibly gas/gas. All of them, though can be classified generally as interfaces, their interface properties can be quite different and hence may require independent treatment. Ideally, it will be good if one can treat all interfaces under one framework, but due to various limitations, we will be confining ourselves mainly to the solid/gas interface though some of the studies that we consider here can be extended to other interfaces like gas/liquid and solid/liquid interfaces as well.

The solid/gas interface is unique in many respects. Many of the chemical industrial processes that are practiced today, like cracking of petroleum crude, hydrogenation of fatty acids, oxidation of organic substrates are all reactions that take place on the surface of a solid catalyst by contacting the vapours of the reactants, even though some heterogeneous reactions are also carried out in solid/liquid interfaces. It appears, generally the principles involved in the surface catalytic reactions whether the substrate is in vapour state or in condensed liquid state to be the same and hence, it is necessary that one considers the adsorption of adsorbates on the surface of the solid by suitable means. Since the surface on an atomic or molecular scale is heterogeneous with each of the sites being coordinatively unsaturated not to the same extent (refer to Fig.1.1 reproduced from Ref.1.) each of these sites can exhibit totally different reactivity.

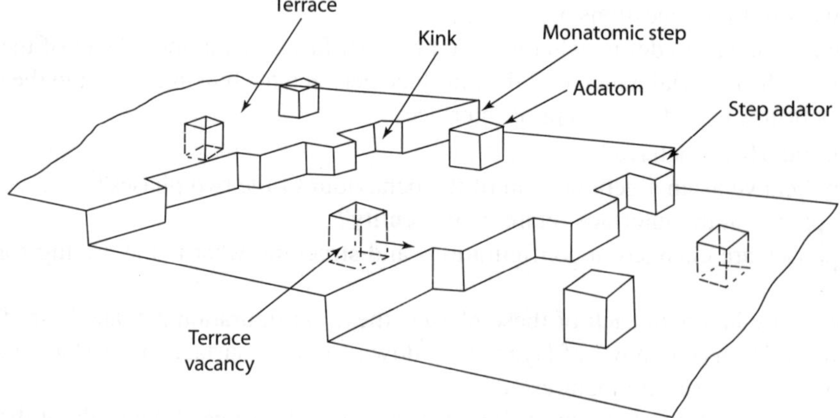

Fig. 1.1 Model of a heterogeneous solid surface, depicting different surface sites. These sites are distinguishable by their number of nearest neighbours

At this stage, one is tempted to name these sites where the reactant molecules are formed into product molecules following the designation of Taylor [2] as the "active sites". It is natural that one will ask what these active sites are?

- ❖ Are they defined in terms of geometry and surrounding coordination or in terms of energetic considerations?
- ❖ Are these active sites *a priori* formed on the surface before the surface comes into contact with the substrate or are they generated at the call of the substrate molecules?
- ❖ Are these so called "active sites" a dynamic concept or a static pictorial representation?

Many of the heterogeneous catalytic phases are usually supported on some (supposedly inert for the substrate) matrix like oxides which do not participate directly in the catalytic reaction. However, one has been encountering situations where the support interacts with the active metallic component and this state has been termed as Metal Support Interaction (MSI). One such interaction has been well studied, termed as Strong Metal Support Interaction (SMSI) originally proposed by Tauster *et al.* [3]. Metal support interactions or as a matter of fact, the combination of two solid phases have been shown to give rise to interesting electronic properties, a concept originally developed by Shawab and his coworkers [4]. This solid/solid interface and the consequent electronic properties have been addressed in terms of hetero junctions for solid state electronic applications. Another important observation that needs our attention is the possibility of the support acting as sink for the active species generated on metal sites at a fast rate compared to the desired catalytic reaction and the re-supply of these active species from the support to metal at a rate comparable to that of the catalytic reaction. This phenomenon is termed as "spill over" by Boudart [5]. Though spillover is considered with respect to hydrogen and oxygen, the concepts developed can be extended to a number of adsorbed species which can be transported to and re-transported from the support with the help of suitable activation species. This means that one has to enlarge the scope of the definition of active sites, since these support oxide sites which act as sink for holding these active species and for their subsequent release, are they active sites or not? It appears that one has to start with the definition of the active site on which the reactant molecules are activated. When a substrate molecule interacts on such an active site, it can do so in a number of ways geometrically and this can result in various modes of adsorption. The possible modes of adsorption of CO and N_2 on typical well defined metallic surfaces are shown in Fig. 1.2.

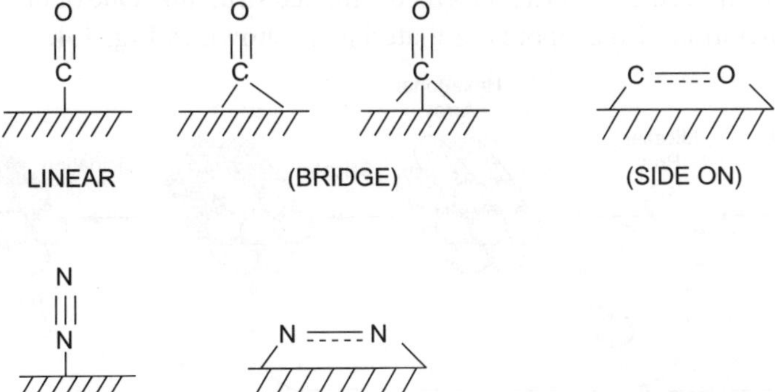

Fig. 1.2 Adsorbed states of CO and N_2 on metal surface

Similarly for many other molecules one can imagine multiple modes of adsorption[6] and the adsorption energy and adsorption geometry may also contribute to the product selectivity one observes in surface catalyzed reaction. Since the adsorption mode and energies decide which bonds are to be broken and which bonds are to be made, activation of substrates by adsorption always leads to product selectivity. However, there are other selectivities that are most often discussed in catalysis by zeolites. They are termed as reactant or product or transition state shape selectivity, in these cases, the selectivity is governed by the shapes of the species formed or taken so as to fit in the zeolite cages and channels.

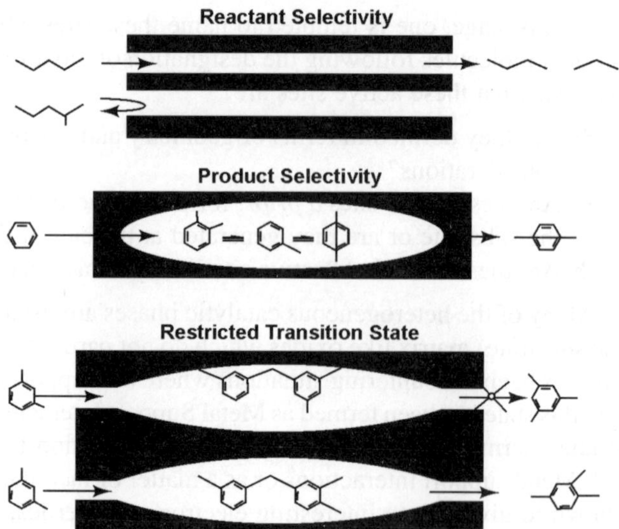

Fig. 1.3 Mechanism of shape selective catalysis

Since the void space in zeolites is similar to the molecular volumes it is necessary that only certain molecules are able to be accommodated or passed through the channels. Others without the particular shape or bigger molecules are trapped inside the cages and undergo transformations till they are able to pass through the zeolite channels. A pictorial representation of these three shape selectivities which are particularly applicable to microporous systems are given in Fig. 1.3. This brings us to the question of the surface area of a given solid. The actual surface area accessible to molecules may be greater than the geometrical area for the given shape of a catalyst. The particles constituting the catalyst pellet or ring or any other preformed shape may arrange themselves randomly so as to give rise to void volumes which are accessible to the reactant and product molecules. Even while forming the solid from suitable precursors like hydroxides, carbonates, sulphates, the elimination of water, CO, CO_2 or SO_2 and their expulsion from the solid oxide mass obtained could give rise to a porous structure.

A third possibility is that the deposition or precipitation of the inorganic substances can take place on individual template molecules or on a preformed template structure. One of the typical mechanisms proposed for the generation of mesopores in materials is shown in Fig. 1.4.

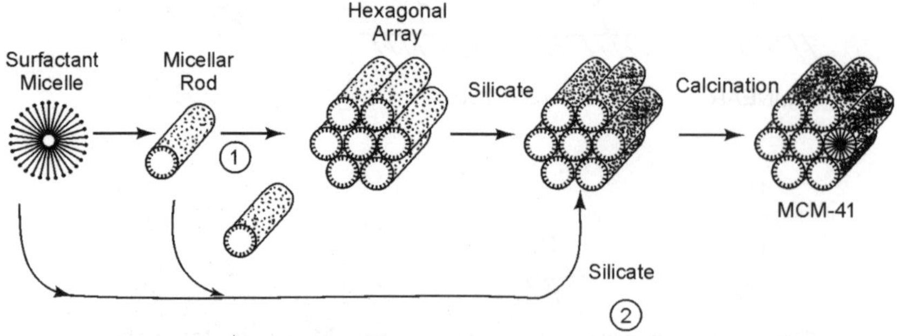

Fig. 1.4(a) Possible mechanistic pathways for the formation of MCM-41 : (1) liquid crystal initiated and (2) silicate anion initiated [7]

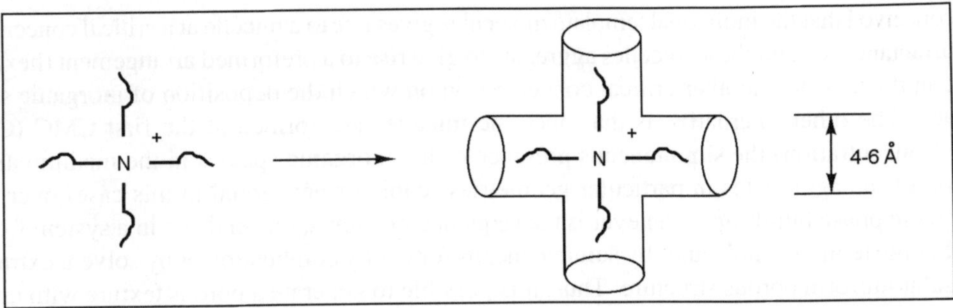

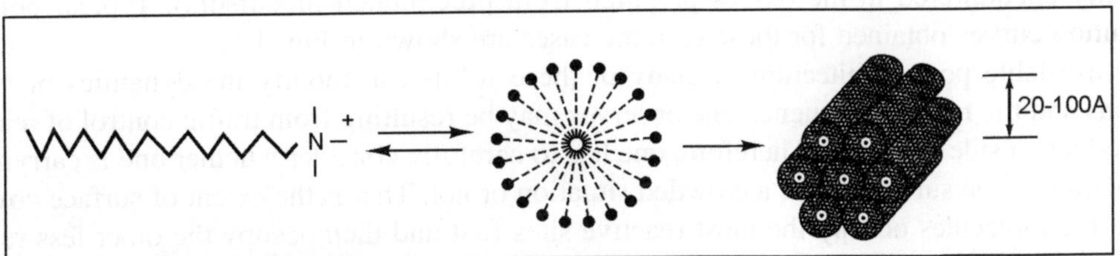

Fig. 1.4(b) Formation of microporous and mesoporous molecular sieves using small (top) and long (bottom) alkyl chain length (quaternary salts) directing agents

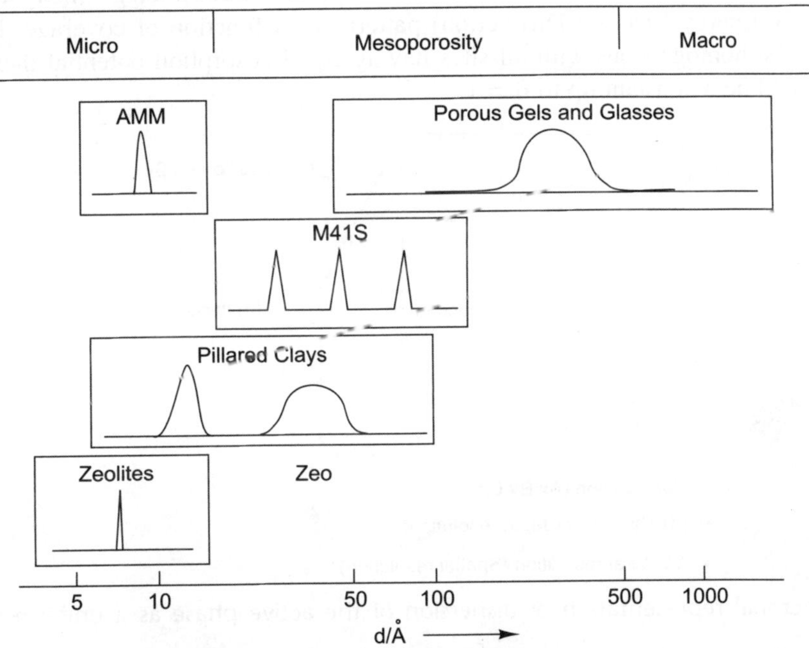

Fig. 1.5 Pore size distribution curves: Porous solid materials are classified according to IUPAC as Microporous d < 20 Å — Zeolites, ALPO's; Mesoporous 20 Å < d < 500 Å — M41S, HMS, MSU ; and Macroporous d > 500 Å — Perovskites, Glasses

It is conceived that the individual template molecules gives rise to a micelle at a critical concentration of the surfactant used and these micelles aggregate to give rise to a preformed arrangement (hexagonal or cubic in this case) at another critical concentration on which the deposition of inorganic species take place. The other alternative is that once the micelles are formed at the first CMC (Critical Micelle Concentration) the simultaneous presence of the inorganic species in the medium aids the formation of the aggregates in particular geometries (cubic or hexagonal in this case) over which the inorganic phase builds up. Whatever is the sequence of events, one ends up in a system from the removal of the template molecules by suitable means (either by combustion or by solvent extraction) with regular uniform porous structure. Thus, it is possible to generate a porous texture with uniform narrow pore size distribution, as is often the case for zeolites or a wide pore size distribution (normally encountered in the oxides prepared from precipitated precursors). Typical pore-size distribution curves obtained for these extreme cases are shown in Fig. 1.5.

The available pore architecture in many of these solids can modify the dynamics of surface reactions and the reaction sequence one observes may be resulting from traffic control of reactants and products inside the pores. Therefore, one has to carefully consider whether one is carrying out the reaction on the surface under a crowded situation or not. That is the extent of surface coverage (θ). As the molecules occupy the most reactive sites first and then occupy the other less reactive sites, the activation of molecules is a function of coverage. As coverage increases (i.e. as $\theta \to 1$) the adsorbed molecules interact among themselves laterally and this interaction is superimposed on the interaction of reactant molecules with the surface atoms or species. The consequence of all these effects is manifested in the variation of the heat of adsorption or adsorption geometry as revealed by the LEED (Low Energy Electron Diffraction) patterns as a function of coverage. Ideally if the surface was to be homogeneous with all sites having equal adsorption potential then the heat of adsorption should be a constant up to $\theta = 1$.

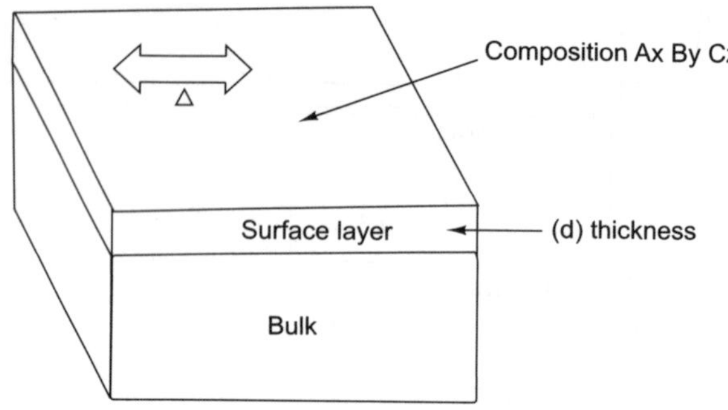

- Composition (Ax By Cz)
- (d) Thickness (depth resolution)
- Δ Lateral resolution (Spatial resolution)

Fig. 1.6 Pictorial representation of dispersion of the active phase as a uniform monolayer

In the case of supported catalysts there are yet other variables like the crystallite sizes and dispersion. For an efficient catalyst the ideal situation would be to disperse the active phase as a uniform monolayer on the surface of the support. Such a situation is shown in Fig. 1.6. This situation is an ideal situation and what one achieves is a thin film consisting of a few layers of the active material on the support phase.

Thin films of this kind have various other important applications and these will not be considered in the present context. When the supported species crystallize and form a crystallite, then the number of atoms exposed will decrease with the total number of particles in the crystallite as well as the shape of the crystallite. One will be formation of clusters of atoms into simple cubic packing 8, 27, 64, 125, 216, 343, 512 atoms. In the 8 atom cluster all the atoms are present on the surface, however, dispersion 'D', defined as the ratio of the number of atoms on the surface to the total number of atoms in the crystallite, decreases rapidly with increase in the total number of atoms in the clusters. The values of dispersion is also dependent on the shape of the cluster formed, and as an example, for a cubo-octahedral arrangement the variation of dispersion as a function of crystallite size is shown in Fig. 1.7.

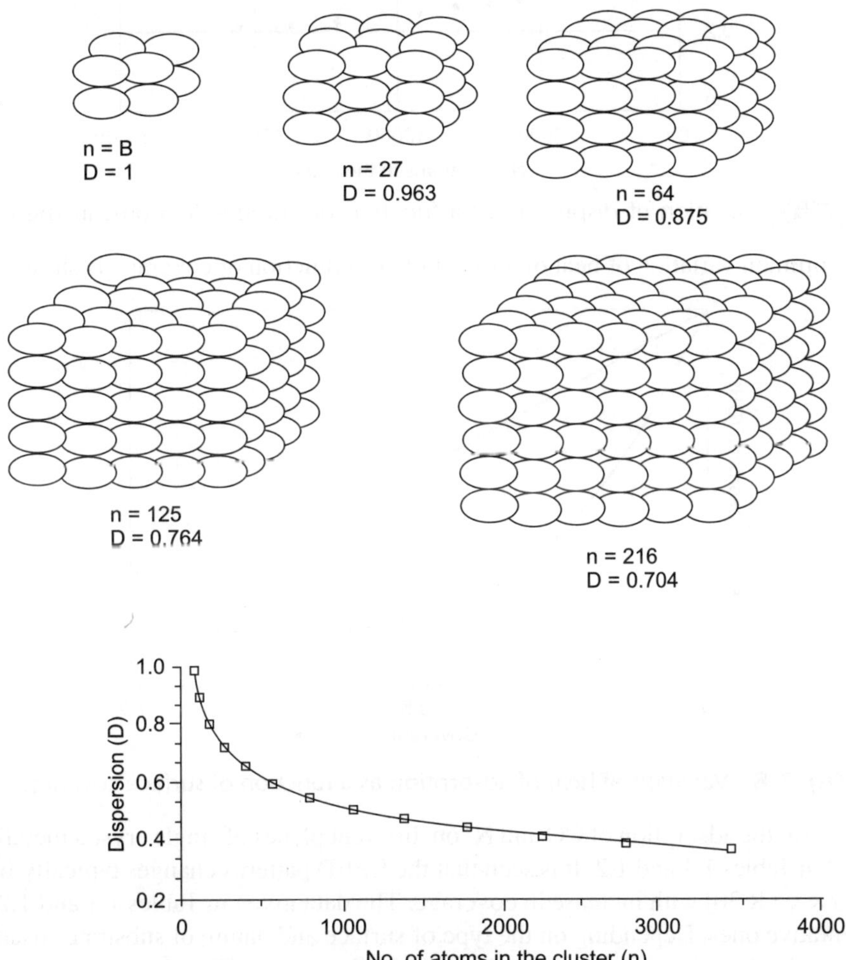

Fig. 1.7(a) Clusters of atoms with single cubic packing having 8, 27, 64, 125 and 216 atoms. In an eight-atom cluster, all of the atoms are on the surface. However, the dispersion 'D', defined as the number of surface atoms divided by the total number of atoms in the cluster, declines rapidly with increasing cluster size, this is shown in the lower part of the figure

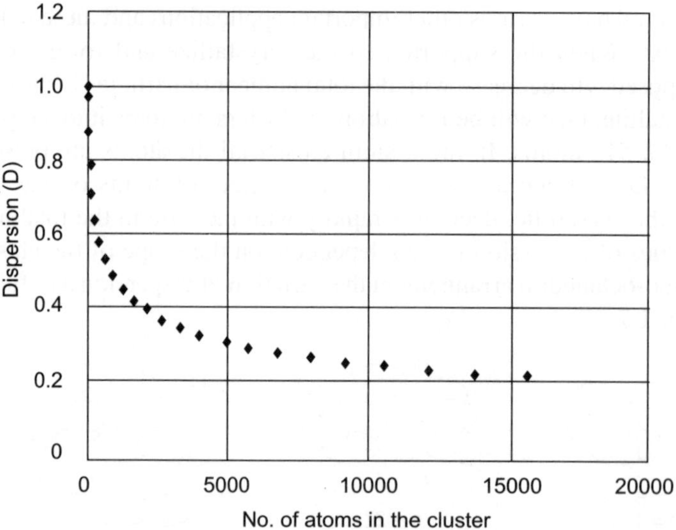

Fig. 1.7(b) Variation of dispersion as a function of number of atoms in the cluster

The most common variation of heat of adsorption as a function of coverage is shown in Fig. 1.8.

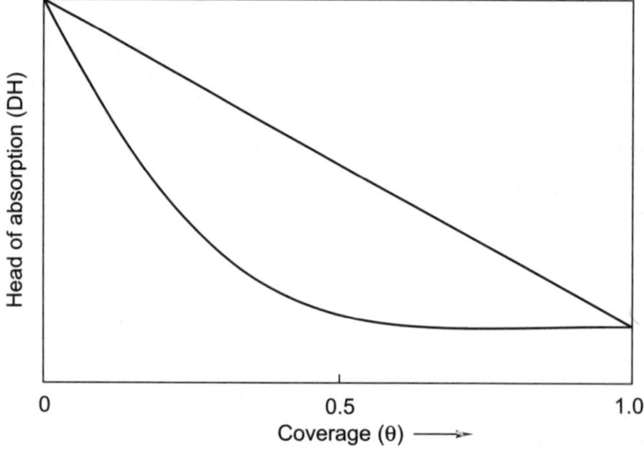

Fig. 1.8 Variation of heat of adsorption as a function of surface coverage

Typical data for the adsorption of CO and N_2 on different planes of single crystal metallic surfaces are assembled in Tables 1.1 and 1.2. It is seen that the LEED patterns changes typically from simple $[1 \times 1]$ to $[\sqrt{3} \times \sqrt{3}\, R\, 30]$ with increase in coverage. The data given in Tables 1.1 and 1.2 are meant to be representative ones. Depending on the type of surface and nature of substrates used, there can be many other adsorbate structures as revealed by LEED patterns. The electronic properties of the adsorbates also change on adsorption. Let us consider as example the adsorption of CO. The molecular orbital diagram in its ground state is shown in Fig. 1.9 and the appropriate molecular properties of neutral CO, CO^+, CO^- and CO^* are given in Table 1.3.

Table 1.1 Parameters for the adsorption of carbon monoxide on single crystal metal surfaces(Data from S. Ishi, Y. Ohno and B. Viswanathan, J. Sci. Ind. Res., **46**, 541 (1987))

Metal surface	Adsorbate structure	E kJ/mol change eV	Work function	N (C-O) cm^{-1}
Fe(110)	C(4×2)	147	1.2	1890
Fe(111)	Diffuse (1×1)		1.6	1800
Ni(100)	C(2×2)	126	1.1	1916
Ni(110)	C(2×2)	126	1.3	1960
Ni(111)	$(\sqrt{3}\times\sqrt{3})$R30	113	1.1	1817
Ru(0001)	$(\sqrt{3}\times\sqrt{3})$R30	126	0.65	1984
Os(0001)	$(\sqrt{3}\times\sqrt{3})$R30	138.6		
Co(0001)	$(\sqrt{3}\times\sqrt{3})$R30		1.35	
Rh(111)	P(2×2)	134		1890
Ir(111)	$(\sqrt{3}\times\sqrt{3})$R30	142.8	0.18	
Pd(111)	$(\sqrt{3}\times\sqrt{3})$R30	142	1.0	1813
Pt(100)	C(2×2)	155.4	0.4	2080
Pt(111)	$(\sqrt{3}\times\sqrt{3})$R30	138.6	0.15	2065

Table 1.2 Typical parameters for the chemisorption of nitrogen on single crystal metal surfaces (Data from S. Ishi, Y. Ohno and B. Viswanathan, J. Sci. Ind. Res., **46**, 541 (1987))

Metal surface	Temperature range	Adsorbate structure	Adsorption energy kJ/mol	Work function change eV	v (N-N) cm^{-1}
Fe(100)		C(2×2)	31.5		
Fe(111)	<100		31.5	0.14	
Ni(100)		C(2×2)			2200
Ni(110)	80–125	(2×1)	<47	0.1	2194
Ni(111)	20–100		20		2186
Cr(110)	100		14		
Ru(0001)	95	$(\sqrt{3}\times\sqrt{3})$R30	31.5		2252
Re(1120)	80		21-29		
W(110)	100		25		
Rh(111)			33.6	−0.15	
Ir(111)				−0.2	
Ir(100)			33.6		
Pt(100)			37.8	−0.3	
Pt(111)			38.6		2230

Table.1.3 Summary of molecular states of CO in neutral, ionic and excited states

	CO	CO$^+$	Co$^-$	CO*		Co*	
configuration	$...(1\pi)^4(5\sigma)^2$	$...(1\pi)^4(5\sigma)^1$	$.....(1\pi)^4$ $(5\sigma)^2(2\pi)^1$	$.....(1\pi)^4$ $(5\sigma)^1(2\pi)^1$		$.....(1\pi)^3$ $(5\sigma)^2(2\pi)^1$	
				$(\uparrow\uparrow)$	$(\uparrow\downarrow)$	$(\uparrow\uparrow)$	$(\uparrow\downarrow)$
State	$^1\Sigma^+$	$^2\Sigma^+$	Π	A $^1\Pi$	A' $^3\Pi$	$^3\Sigma^+$	I' $^2\Sigma^+$
R_e A	1.13	1.12	1.17	1.21	1.24	1.35	1.39
N (CO) cm^{-1}	2143	2214	~1850	1743	1518	1229	1092
D (eV)	11.2	8.4	8.1	5.21	3.17	4.32	3.17
Ì(Debye)	0.1 ($^-$CO$^+$)			1.38 $^+$CO$^-$		1.06 $^-$CO$^+$	
I (eV)	14.01	26.8	-1.5				

Reproduced from S. Ishi, Y. Ohno and B. Viswanathan, Surface Sci., 161, 349(1985).

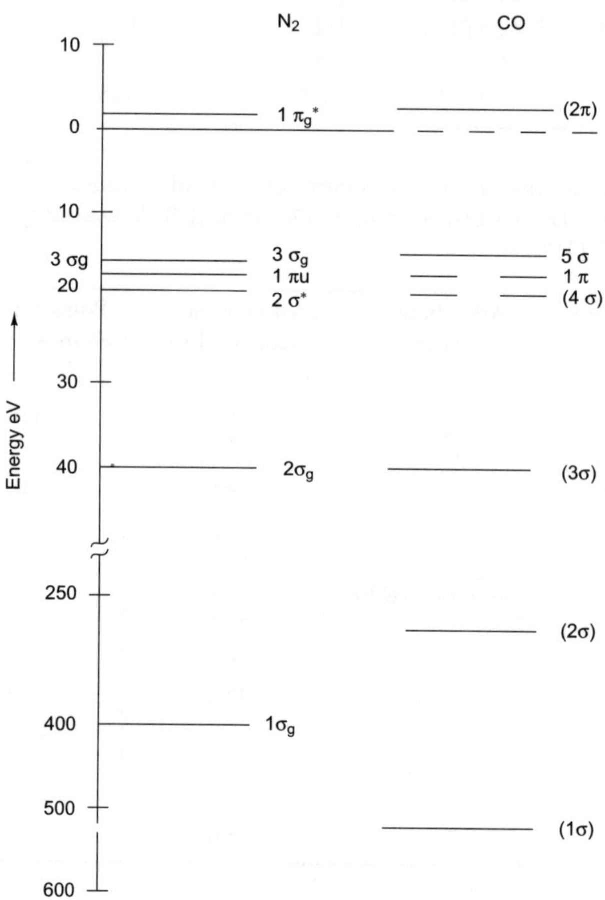

Fig. 1.9 The MO diagram of the ground state N$_2$ and CO

It is seen that the highest occupied states in free CO molecule are 5σ and 1π in this order. However, on adsorption the ordering 5σ and 1π is inversed or the photoemission from these two states overlap meaning that 1π level is shifted to lower binding energies or the energy required for ionization from 5σ level has increased. This aspect is normally considered in terms of (4σ - 1π), the separation between 4σ and 1π levels. The relevant data for this $|4\sigma$ - $1\pi|$ separation between free molecules as well as for CO in the adsorbed state are assembled in Table 1.4.

It is to be recognized that the position of the various energy levels in the free molecule is altered as a result of adsorption. This alteration can be explained with respect to the unoccupied LUMO state. This aspect has to be probed by techniques which can monitor unoccupied states like 'inverse photo emission' or by probing the satellite features of the core levels of CO in the adsorbed state. Such an experimental data for emission from 1σ and 2σ levels of CO are given in Fig. 1.10 and the corresponding data extracted from these figures are given in Table 1.5.

Table 1.4 Peak positions of valence levels with respect to E_F and energy separation for $4\tilde{\sigma}$ and $1\tilde{\pi}$ and $(\Delta|(4\tilde{\sigma} - 1\tilde{\pi})|)$ for CO adsorption on group VIII metals

	Peak positions eV				
	$5\tilde{\sigma}$	Overlapping 5σ and 1π	$1\tilde{\pi}$	$4\tilde{\sigma}$	$(\Delta|(4\tilde{\sigma} - 1\tilde{\pi})|)$
Fe(110)	8.1		6.9	11.0	4.1
Fe (100)		7.0		10.5	3.5
Fe(111)		7.6		10.8	3.2
Ru(0001)		7.6		10.7	3.1
Co(0001)		7.5		11.0	3.5
	7.9		7.7	10.7	3.0
Rh(111)	8.3		7.8	11.2	3.4
Rh(100)		7.6		10.8	3.2
Rh(110)	7.6		8.7	10.6	1.9
Ir(111)	9.2		8.6	11.7	3.1
Ir(100)	9.2		8.6	11.7	3.1
Ni(111)	8.1		7.1	11.2	4.1
Ni(100)	8.3		7.8	10.8	3.0
Ni(110)	8.4		6.5	11.7	5.2
Pd(111)	8.2		7.5	11.2	3.7
Pd(100)		7.9		10.8	2.9
Pd(110)		7.		10.8	2.9
Pt(111)	9.2		8.4	11.9	3.5
Pt(100)		8.8		11.4	2.6
Pt(110)	9.2		8.2	11.7	3.5
CO (gas)	14.0		16.9	19.7	2.8

Data extracted from S. Ishi, Y. Ohno and B. Viswanathan, Surface Sci., 161, 349(1985)

Table 1.5 Positions of shake-up features in adsorbed CO or N$_2$ on Ni(100)

C1s (eV)	O1s (eV)	Inner Nitrogen (1s) (eV)	Outer Nitrogen (1s) (eV)
2.1 (sh)	5.5		2.1
5.5	8.0 (sh)	5.3	5.8
9.5	15.0		8.5
		~15	~15.0
	16.0		
33 (σ to nσ*)	26.0		
	35.0		
	45.0		
	55.0		

Data extracted from Anders Nilsson and Nils Mårtensson, Phys. Rev. B 40, 10249 - 10261 (1989); Anders Nilsson, Hlena Tillborg and Nils Mårtensson, Phys. Rev. Lett., 67, 1015(1991)

The satellite features appear because of the excitations from 5σ to 2π* or 1π to 2π* levels and the energy loss of photo emitted electrons with respect to the main emission denotes the 5σ → 2π* or 1π → 2π* separation values which have to be compared with the corresponding values of free gas phase CO. The consequence of the perturbations of the electronic energy levels of molecules in the adsorbed state is manifested in the vibrational properties of these molecules in the adsorbed state. The free gas phase CO molecule gives rise to an IR absorption band at 2143 cm^{-1}. This band is shifted to lower values of the wave number indicating the weakening of the C-O bond as a result of adsorption.

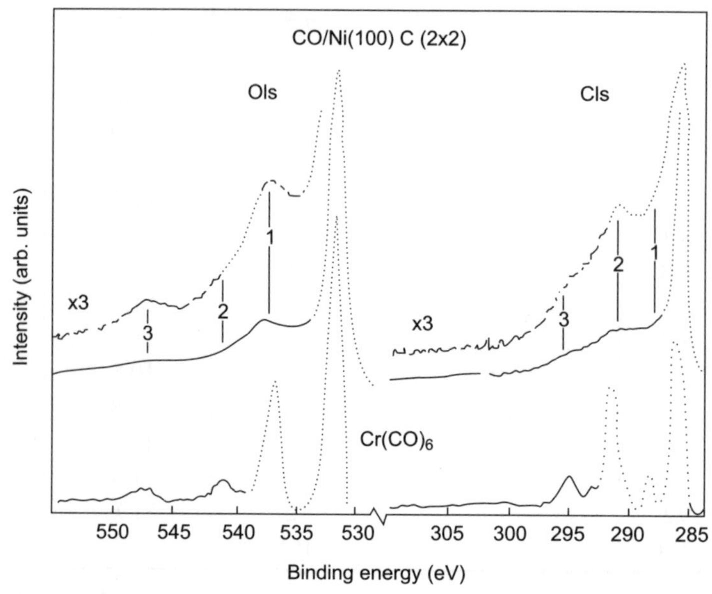

Fig. 1.10 C 1s and O 1s XP spectra for CO/Ni(100) c(2x2) and Cr(CO)$_6$ (Suitable background correction to the adsorbate spectra has been provided for and the resultant absorbate spectra alone are shown)

Table 1.6 Typical IR frequencies of adsorbed CO on single crystal metal surfaces

surface	Peak positions (cm^{-1})			
	v(C-O)		v(M-CO)	
	$\Theta = 0$	$\Theta \geq 0.5$	$\Theta = 0$	$\Theta \geq 0.5$
Fe(110)	1890	1985	456	360,440
Fe(111)	1800	1850, 2000	550	515
Ru(0001)	1984	2060	445	445
Rh(111)	1990	2070, 1870	480	420
Ni(111)	1817	1915, 2045	380	380
Ni(100)	2000	2070		480, 365
Ni(110)	1990, 1880	2015, 1935		450, 345
Pd(111)	1813	1936, 2092		
Pd(100)	1985	1950, 2096	347	315
Pt(111)	2065	2100.1872	498	365
Pt(100)	2080, 1950	2096		
Cu(111)	2078	2070		
Cu(100)	2079	2088		
Cu(110)	2088, 2104	2094		

Extracted from S. Ishi, Y. Ohno and B. Viswanathan, Surface Sci., 161, 349(1985).

A compilation of the position of the IR absorption bands of adsorbed CO on various metal surfaces is given in Table 1.6. Though one can generally consider that greater is the shift of this IR band to lower values with respect to the value of 2143 cm^{-1} corresponding to gas phase CO, (Δv) the weaker is the C–O bond. This statement holds well only when CO is adsorbed on top mode. If CO was to be adsorbed in bridged configuration as bidentate or tridentate species (See structures in Fig. 1.2.) then also one would observe the IR bands at lower frequency. Some typical data for the position of IR bands for CO bonded in various configurations are given in Table 1.7.

Table 1.7 Typical IR frequencies for various types of bonded CO

Type of Co bonded	IR Frequency cm^{-1}
Uncoordinated or "free" CO	2143
Terminal M-CO	2125 to 1850
Doubly bridging (mu-2)	1850 to 1750
Triply bridging (mu-3)	1675 to 1600
Semi-bridging	between terminal and mu^{-2}.

Normally the (coverage) number of adsorbed molecules per unit area (σ) (cm^2) is given by $\sigma = F \times \tau$, Where F is the incidence flux of molecules cm^{-2} sec^{-1} and τ is the surface residence time

$$\tau = \tau_0 \exp [\Delta H_{ads} (\theta)/RT]$$

Where τ_0 is typically 10^{-13} sec. The adsorbate-adsorbate interaction decides the magnitude of ΔH and the incidence flux is given by

$$F = P \, (2\pi \, MRT)^{-1/2}$$

While on the point of selectivity, one wishes to consider varying levels of selectivity and the selectivity order that is currently being pursued by chemists is given in Fig. 1.11.

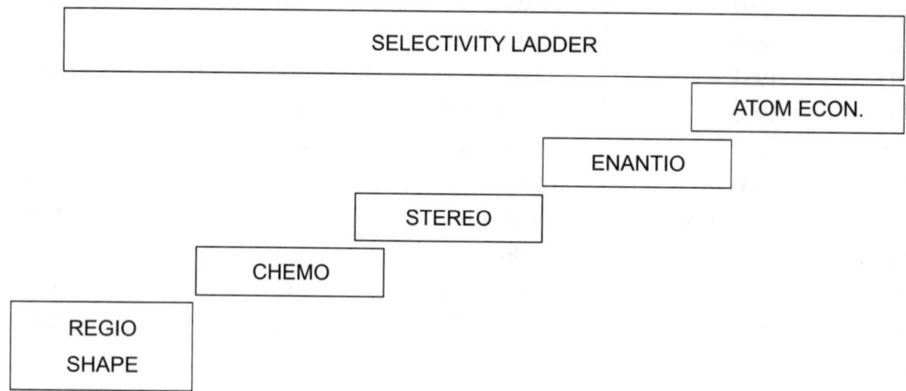

Fig. 1.11   Selectivity ladder presumed by chemists (www.sc.doe.gov/bes/BESAC/ Catalysis Marks 05-15-02.ppt)

Adsorption generally means the residence of the molecules at the surface of a solid. With in its residence time, the bond activation takes place so as to provide bond transformation in the desired directions. The reaction yield can be represented in terms of Turn Over Number (TON) or Turn Over Frequency (TOF). TON is defined as specific rate multiplied by the time for which the reaction is carried out. Specific rate denotes the number of molecules converted per unit site per second at a given pressure, concentration, temperature conditions. This is a figure of merit of the activities of catalysts for comparison purposes and it can range from 10^{-4} to 10 per site per second. Similarly, TON's must be greater than 1, for catalytic reactions and it will be less than 1 when one is dealing with stoichiometric reactions. Long residence time of molecules on the surface may lead to migration of adsorbed species (if surface diffusion is a feasible process) over long distances on the surface 10 to 10^4 Å, wherein these adsorbed molecules may meet a number of catalytically active sites where they can undergo consecutive reactions or molecular rearrangements and thus reducing the desired selectivity.

Though the ideal situation would be total atom economy, wherein the total number of atoms that were introduced as reactant should appear in the desired product, such a realization of all the atoms of the reactant molecules being present in the product molecule implies that the reaction is 100% selective without the formation of any product arising out of side reactions. We seem to have most of these types of selectivity realizable in catalytic reactions.

It is appropriate that we have some time line for the developments in the field of surface science and catalysis. One such conceptual time line is shown in Fig. 1.12.

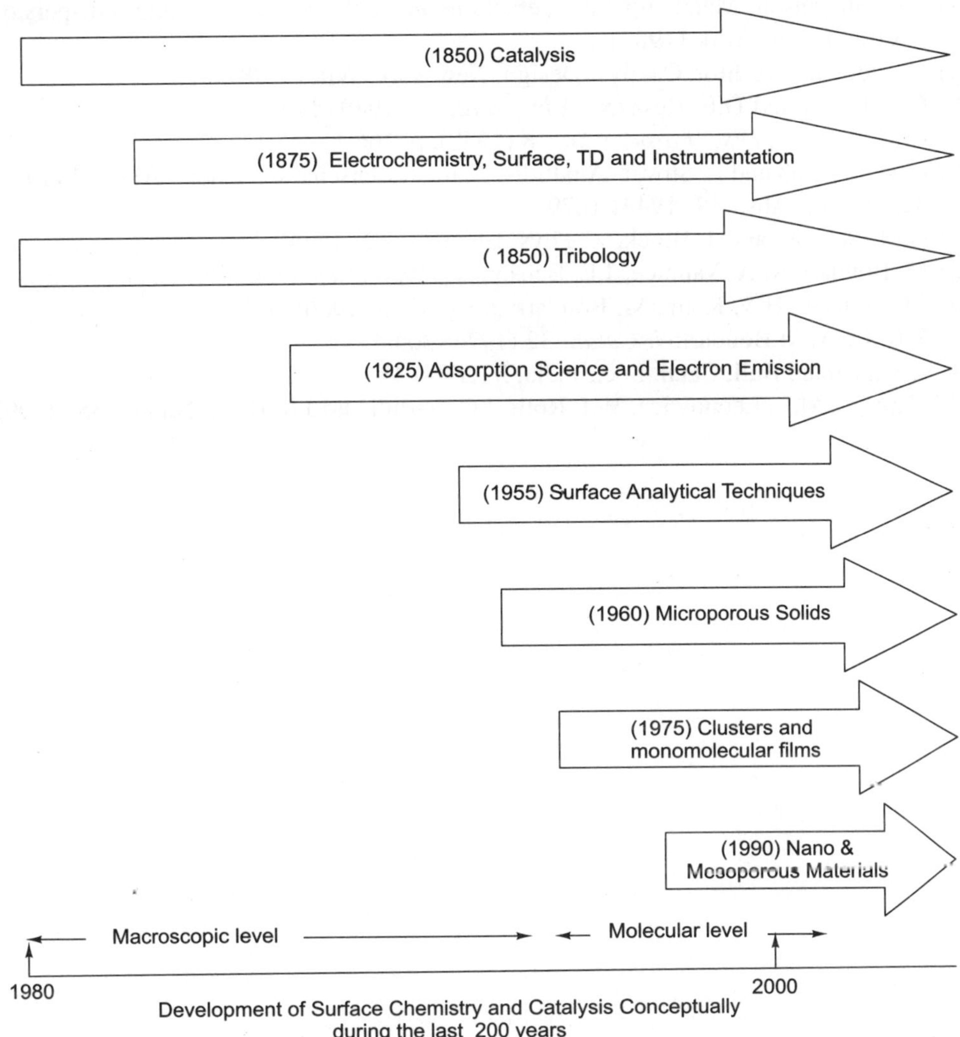

Fig. 1.12   Conceptual time line for the development in the Field of Surface Science and Catalysis

REFERENCES

1. G.A. Somorjai,, Chemistry in Two Dimensions: Surfaces, Cornell University Press, 1981
2. H.S. Taylor. *Proc. R. Soc. Lond. A* 108 (1925) 105
3. (a) S.J. Tauster, S.C. Fung and R.L. Garten. *J. Amer. Chem. Soc.* 100 (1978) 170
   (b) S.J. Tauster and S.C. Fung, *J. Catal* 55 (1978) 29
   (c) S.J. Tauster, S.C. Fung, R.T.K. Baker and J.A. Horsley, *Science* 217 (1981) 1121
   (d) E.C. Akubuiro and X.E. Verykios, *J. Catal* 103 (1987) 320
   (e) E.C. Akubuiro and X.E. Verykios, *J. Phys. Chem. Solids* 50 (1989) 17

(f) S.A. Stevenson, Metal-support interactions in catalysis, sintering and redispersion, Van Nostrand, New York (1987)

(g) Hegedus L.L., Editor, Catalyst Design. New York: Wiley, 1987

(h) G.L. Haller and D.E. Resasco, *Adv. Catal.* 36 (1989) 173

4. (a) G.M. Schwab, *Disc. Faraday Soc.* 8 (1950), p. 166

(b) G.M. Schwab and E. Shwab-Agallidis, Naturwissenschaften, Chem. Abs., 31 (1943) 322-323; Chem. Abs., 38 (1944) 1679

(c) G.M. Schwab and J. Block, Z. Phys. Chem. 1, 42 (1954)

5. (a) M. Boudart, M.A. Vannice, J.E. Benson: *J. Phys. Chem.*, 64,(1969) 171

(b) J.E. Benson, H.V. Kohn, M. Boudart: *J. Catal.*, 5 (1966) 307

(c) R.B. Levy, M.Boudart: *J. Catal.*, 32 (1974) 304

6. B. Viswanathan, Bull. Catal. Soc., India, 3 (2004) 43

7. C.T. Kresge, M.E. Leonowicz, W.J. Roth, J.C. Vartuli and J.S. Beck, Nature, 359 (1992) 710

Surface Analytical Techniques: An Overview

The analysis of surface starts with defining what a surface is? Generally, it is considered that the top 4 – 5 layers or top 10″Å depth is considered to be the surface since coordinative unsaturation and variations in configurations can be seen only in these layers. It is usual to list out the available techniques in tabular form. However, since this will not give the whole picture, a method based on input/output probes has been adopted. There are generally four particle beams, namely electrons, ions, neutrals and photons and there are four other fields, namely thermal, electrical, magnetic and surface sonic waves that can be used as input probes. When one employs any one of these eight probes, they give rise to emission or transmission or scattering of the four particle beams (except the magnetic field) namely electrons, ions, neutrals and photons. These particles carry information of the surface to a suitable detector. The detector assembly can be tuned to count the number of particles emitted (intensity), or it can be tuned to identify the chemical nature of the species emitted in the case of ions and neutrals or can be made to analyse the energy or angular distribution of the particles emitted. Any or all of these four types of information on the emitted particles are used to develop better understanding of the surface under study.

The combination of the 8 different types of input probes with four different output probes on which four different types of information can be gathered give rise to the multitude of techniques. A pictorial representation of this model of generating all the techniques is given in Fig. 2.1. A listing of the possible techniques and their acronyms are given in Appendix 2.1.

It can be demonstrated that most of the surface analytical techniques can be rationalized in terms of input and output probes. Consider the case of electrons-in (electrons used as input probe) which can give rise to all the four particle beams. It is generally believed that it is always simple and easily feasible if the same particle is considered as the output probe.

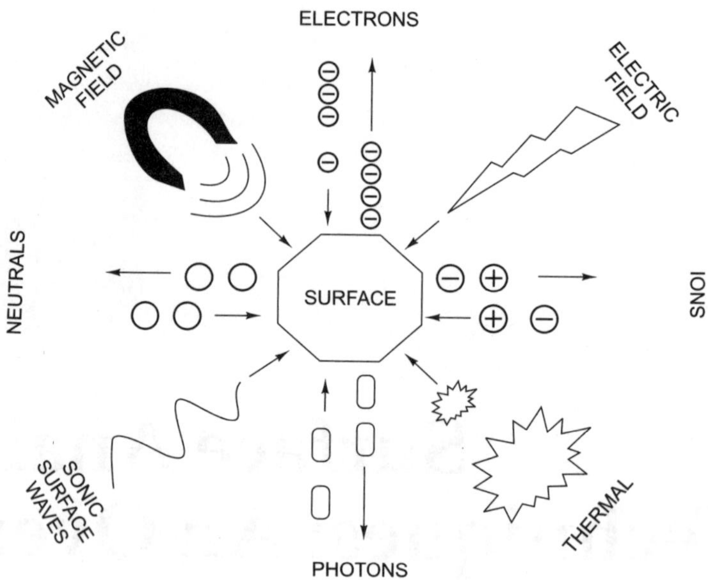

Fig. 2.1 A pictorial representation of generating the possible techniques in terms of input and output probes

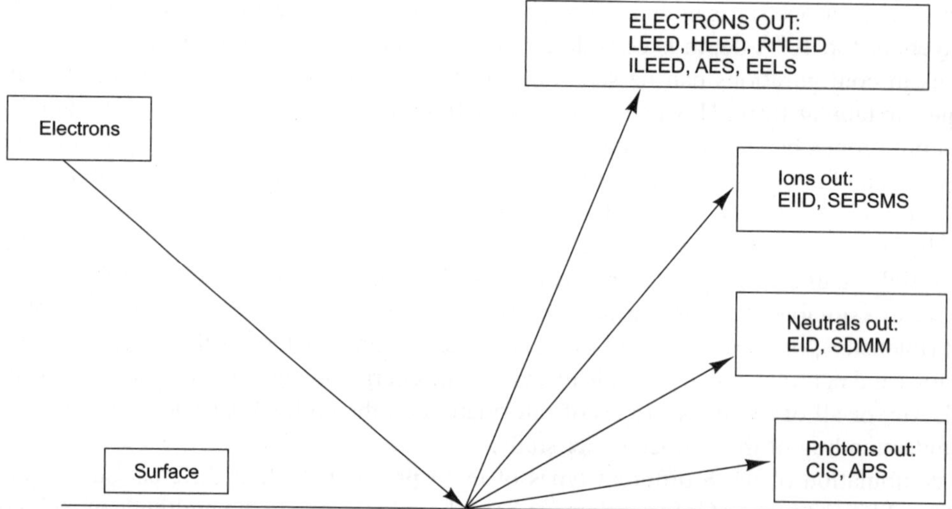

Fig. 2.2 Representation of the techniques based on Electrons in – electron, ion, neutral and photon out **LEED:** Low Energy Electron Diffraction; **HEED:** High Energy Electron Diffraction; **RHHED:** Reflected High Energy Electron Diffraction; **ILEED:** Ineleastic Low Energy Electron Diffraction; **AES:** Auger Electron Spectroscopy; **EELS:** Electron Energy Loss Spectroscopy; **EIID:** Electron Induced Ion Desorption; **SEPSMS:** Electron Probe Surface Mass Spectrometry; **EID:** Electron Induced Desorption; **SDMM:** Surface Desorption Molecular Microscope; **CIS:** Characteristic Isochromat Spectroscopy; **APS:** Appearance Potential Spectroscopy

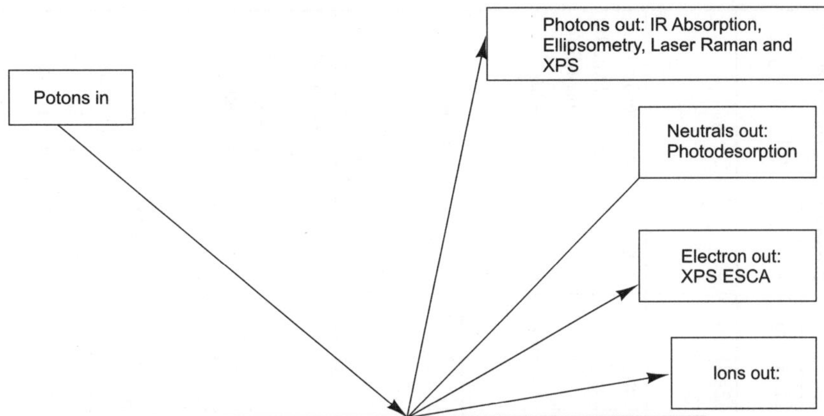

Fig. 2.3   Schematic representation of the techniques that can be generated from Photon-in photon, neutral, electron or ion-out methodology. **XPS:** X ray Photoelectron Spectroscopy; **ESCA:** Electrons Spectroscopy for Chemical Analysis

For a conceptual understanding a simple representation is given in Fig. 2.2 for the surface analytical techniques that are possible from using electrons as input probes and all the four particle beams as out put probes. A similar representation is possible for other combinations as well. Some of these are pictorially shown in Fig. 2.3 and Fig. 2.4. A simple compilation of some of the important surface analytical techniques, their basis and the type of information that can be obtained is given in Table 2.1.

The purpose of this presentation is not to consider the fundamentals of these techniques as several authoritative compilations are available on them. The focus will be on the applications of some of these techniques in handling the problems in catalysis

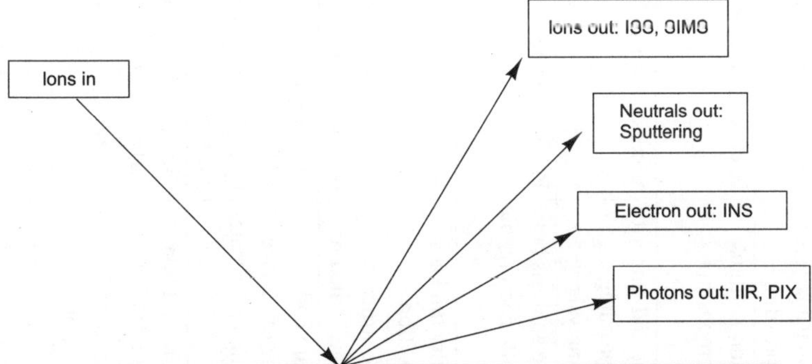

Fig. 2.4   Schematic representation of the techniques that can be generated from Ions-in ion-, neutral-, electron- or photon-out methodology. **ISS:** Ion Scattering Spectroscopy, **SIMS:** Secondary Ion Mass Spectrometry, **INS:** Ion Neutralization Spectroscopy, **PIX:** Proton Induced X ray emission

It is desirable to deal in detail with the various applications of the surface and bulk analytical techniques and show in which situation, which technique will provide the appropriate information required. However, such an exercise will be out of reach for any single volume. Hence in this volume only a few selected techniques are taken up for consideration. Even in these cases the examples chosen do not cover the entire spectrum of possibilities.

Table 2.1 Typical information that can be obtained employing surface analytical techniques and the possible limitations of these techniques

Surface Analytical technique	Typical applications	Signal detected	Elements detected	Detection limits	Depth resolution	Imaging/ Mapping possibility	Lateral resolution (Probe size)
Auger spectroscopy	Elemental analysis, depth profiling	Atomic scale, roughness	Li-U	-	2-20 nm	yes	100 nm
Rutherford Back Scattering (RBS)	Quantitative think film composition	Backscattered He atoms	Li-U	1-10 at% (for Z<20)0.01-1 at % for X-20-70	2-20 nm	yes	2 mm
Secondary Ion Mass Spectrometry	Dopant and impurity depth profiling, microanalysis	Secondary ions	H-U	ppb/ppm	<5 nm	yes	<5 micron ima-ging <30 micron depth profiling
X-ray Photo-electron Spectroscopy	Surface analysis both inorganic and organic	Photoelectrons	Li-U	0.01-1 at%	1-10 nm	yes	10μm-2μ
X ray Fluorescence	Thin film thickness composition	X-rays	Na-U	10 ppm	-	no	100μm
Low Energy Elec-tron Diffraction	Surface structure adsorbate structure	Elastic back scattering of low energy electrons	Only geometry	submonolayer	-	yes	Atomic dimensions
High Resolution Electron Energy Loss Spectroscopy	Structure and bonding of surface atoms and adsorbates	Vibrational excitation of surface atoms adsorbates by inelastic low energy electrons	All adsorbate molecuels	Sub monolayer	-	-	Observation of direct adsorbate-adsorbent bond
Infra red absorption spectroscopy	Structure and bon-ding of adsorbates	Vibrational excitation of surface bonds	Adsorbates internal bonds	Sub monolayer	-	-	-
Ion Scattering Spectroscopy	Atomic structure composition	Elastic reflection of inert gas ions	Any element mass dependent	-	possible	possible	Atomic dimensions
Extended X-ray Absorption Fine structure	Atomic structure of surface atoms and adsorbates	Interference effects in photo-emitted electron wave function in X-ray absorption.	Mostly all species	Intermediate-coordination	-	possible	Atomic dimensions
Thermal Desorption Spectroscopy	Adsorption energy	Thermally induced desorption or decompo-sition of adsorbates	All species	Sub monolayer	Not normally done	-	No

Appendix 2.1

Partial listing of the surface Analytical Techniques with the appropriate acronym.

Acronym	Technique
AEAPS	Auger Electron Appearance Potential Spectroscopy
AES	Auger Electron Spectroscopy
AFM	Atomic Force Microscopy
APECS	Auger Photoelectron Coincidence Spectroscopy
APFIM	Atom Probe Field Ion Microscopy
APS	Appearance Potential Spectroscopy
ARPES	Angle Resolved Photoelectron Spectroscopy
ARUPS	Angle Resolved Ultraviolet Photoelectron Spectroscopy
ATR	Attenuated Total Reflection
BEEM	Ballistic Electron Emission Microscopy
BIS	Bremsstrahlung Isochromat Spectroscopy
CFM	Chemical Force Microscopy
CHA	Concentric Hemispherical Analyzer
CMA	Cylindrical Mirror Analyzer
CPD	Contact Potential Difference
CVD	Chemical Vapour Deposition
DAFS	Diffraction Anomalous Fine Structure
DAPS	Disappearance Potential Spectroscopy
DRIFT	Diffuse Reflectance Infra-Red Fourier Transform
EAPFS	Extended Appearance Potential Fine Structure
EDX	Energy Dispersive X-ray Analysis
EELS	Electron Energy Loss Spectroscopy
E/L	Electron Induced Luminescence
	Ellipsometry
EMS	Electron Momentum Spectroscopy
EPMA	Electron Probe Micro-Analysis
ESDMS	Electron Stimulated Desorption Mass Spectrometry
ESCA	Electron Spectroscopy for Chemical Analysis
ESD	Electron Stimulated Desorption
ESDIAD	Electron Stimulated Desorption Ion Angle Distributions
EXAFS	Extended X-ray Absorption Fine Structure
FEM	Field Emission Microscopy
FIM	Field Ion Microscopy
FTIR	Fourier Transform Infra Red
FTRA-IR	Fourier Transform Reflectance Absorbtion-Infra Red
HAS	Helium Atom Scattering
HDA	Hemispherical Deflection Analyser
HEIS	High Energy Ion Scattering
HREELS	High Resolution Electron Energy Loss Spectroscopy
IETS	Inelastic Electron Tunnelling Spectroscopy

(Contd.)

Acronym	Technique
KRIPES	K-Resolved Inverse Photoemission Spectroscopy
ILS	Ionisation Loss Spectroscopy
INS	Ion Neutralisation Spectroscopy
IPES	Inverse Photo Emission Spectroscopy
IRAS	Infra-Red Absorption Spectroscopy
	Ion Excited Auger Electron Spectroscopy
ISS	Ion Scattering Spectroscopy
LEED	Low Energy Electron Diffraction
LEEM	Low Energy Electron Microscopy
LEIS	Low Energy Ion Scattering
LFM	Lateral Force Microscopy
MBE	Molecular Beam Epitaxy
MBS	Molecular Beam Scattering
MCXD	Magnetic Circular X-ray Dichroism
MEIS	Medium Energy Ion Scattering
MFM	Magnetic Force Microscopy
MIES	Metastable Impact Electron Spectroscopy
MIR	Multiple Internal Reflection
MOCVD	Metal Organic Chemical Vapour Deposition
MOKE	Magneto-Optic Kerr Effect
NEOM	Near Field Optical Microscopy
NIXSW	Normal Incidence X-ray Standing Wave
NEXAFS	Near-Edge X-ray Absorption Fine Structure
NSOM	Near Field Scanning Optical Microscopy
PAES	Positron Annihilation Auger Electron Spectroscopy
PECVD	Plasma Enhanced Chemical Vapour Deposition
PEEM	Photo Emission Electron Microscopy
PhD	Photoelectron Diffraction
	Photon Excited Auger Electron Spectroscopy
PIXE	Proton Induced X-ray Emission
	Raman Spectroscopy
PSD	Photon Stimulated Desorption
RAIRS	Reflection Absorbtion Infra-Red Spectroscopy
RAS	Reflectance Anisotropy Spectroscopy
RBS	Rutherford Back Scattering
RDS	Reflectance Difference Spectroscopy
REXAFS	Reflection Extended X-ray Absorption Fine Structure
RFA	Retarding Field Analyser
RHEED	Reflection High Energy Electron Diffraction
RIFS	Reflectometric Interference Spectroscopy
SAM	Scanning Auger Microscopy
SEM	Scanning Electron Microscopy
SEMPA	Scanning Electron Microscopy with Polarization Analysis

(Contd.)

Acronym	Technique
SERS	Surface Enhanced Raman Scattering
SEXAFS	Surface Extended X-ray Absorption Spectroscopy
SHG	Second Harmonic Generation
SH-MOKE	Second Harmonic Magneto-Optic Kerr Effect
SIMS	Secondary Ion Mass Spectrometry
SKS	Scanning Kinetic Spectroscopy
SMOKE	Surface Magneto-Optic Kerr Effect
SNMS	Sputtered Neutral Mass Spectrometry
SNOM	Scanning Near Field Optical Microscopy
SPIPES	Spin Polarised Inverse Photoemission Spectroscopy
SPEELS	Spin Polarised Electron Energy Loss Spectroscopy
SPLEED	Spin Polarised Low Energy Electron Diffraction
SPM	Scanning Probe Microscopy
SPR	Surface Plasmon Resonance
SPURS	Spin Polarised Ultraviolet Photoelectron Spectroscopy
SPXPS	Spin Polarised X-ray Photoelectron Spectroscopy
STM	Scanning Tunnelling Microscopy
SXAPS	Soft X-ray Appearance Potential Spectroscopy
SXRD	Surface X-ray Diffraction
	Synchrotron Radiation Photoelectron Spectroscopy
TDS	Thermal Desorption Spectroscopy
TEAS	Thermal Energy Atom Scattering
TIRF	Total Internal Reflectance Fluorescence
TPD	Temperature Programmed Desorption
TPO	Temperature Programmed Oxidation
TPRS	Temperature Programmed Reaction Spectroscopy
TPS	Temperature Programmed Sulphidation
TXRF	Total Reflection X-ray Fluorescence
UHV	Ultra High Vacuum
UPS	Ultraviolet Photoemission Spectroscopy
XANES	X-ray Absorption Near-Edge Structure
XAES	X-ray Induced Auger Electron Spectroscopy
XPD	X-ray Photoelectron Diffraction
XPS	X-ray Photoemission Spectroscopy
XRR	X-ray Reflectometry
XSW	X-ray Standing Wave

This compilation is mostly reproduced from http://www.uksaf.org/tech/list.html.

Thermal Methods

Among the various surface analytical techniques thermal field in, neutrals out is the simplest. This technique has been practised for a long time as flash desorption; wherein the temperature of a metallic wire containing adsorbed species is suddenly increased and the desorbed gases analyzed. The principle of thermal desorption is simple. Gaseous molecules are adsorbed on surfaces with various binding energies. If the solid surface temperature is low, due to the binding energy of the adsorbate, these molecules will remain in the adsorbed state on the surface. At room temperature, this situation will prevail for most of the chemisorbed species while at low temperatures such as that liquid nitrogen (77 K) physisorbed species can be retained on the surface. If the surface temperature is then increased, the adsorbed components will desorb at a rate determined by their binding energies, frequency factors, the order of the desorption process and the heating rate of the substrate. The temperature of the surface is increased in a programmed manner either linearly ($T = T_0 + \beta t$ where β is the rate of increase of temperature and t is time in minutes) or exponentially. The linear heating rate has given rise to a number of useful tools including thermogravimetry (TGA), Differential Thermal Analysis (DTA), Differential scanning calorimetry (DSC), as well as a number of temperature programmed techniques usually designated as Temperature Programmed Desorption (TPD), Temperature programmed oxidation (TPO), Temperature Programmed Reduction (TPR), Temperature Programmed Reaction Study (TPRS), and Temperature Programmed sulphidation (TPS).

Generally, temperature programming is better than flash methods in the fact that time and temperature are interrelated in programming techniques and hence one can deduce suitable kinetic expressions for the desorption process or for the process taking place as a result of temperature programming. Temperature programmed techniques are advantageous over isothermal methods on many counts. For example,

1. Detailed stepwise reduction of a metal ion can be examined in TPR while isothermal reduction can give only the overall reduction characteristics.

2. Details of the reduction characteristics of systems containing multi-component metal ions could be probed by temperature programmed technique, while it is difficult to discriminate the reduction schedules of individual ions with isothermal methods.
3. Surface states of metals bound differently with the support can be identified by TPR which is not possible by isothermal techniques.
4. Effect of added promoters, impurities can be studied by the shifts in the reduction maximum or by the appearance of a new maximum in TPR.
5. Peak maximum temperature is a measure of the reactivity or binding strength of the species held on the surface.
6. The integral of the peak is proportional to the number of species involved in reduction, desorption, and re-oxidation.
7. The shifts in the peak maximum temperature in TPD as a function of coverage are indicative of the reaction order.
8. It may be deduced that if reaction products were to evolve at the same peak maximum temperature, then they must originate from the same rate determining step.

In essence, temperature programming appears to provide a handle to discriminate the various individual steps involved in desorption, reduction or oxidation processes while isothermal methods can provide information on the over all process taking place.

Temperature programmed techniques can be advantageously utilized for a number of studies. A kind of listing of these possible utilities are given and these utilities are examined with suitable examples.

1. Mechanistic investigations of the reaction under study.
2. Nature of active sites involved in the reaction.
3. Influence of preparation procedures of the catalyst precursors used on the phases obtained for use subsequently as catalysts.
4. Effect of the support for promoting the formation of active catalyst phase, the chemical composition of the active phase in multi-component systems, the role of promoters for the effective dispersion of the active phase.
5. The interaction between the active phase and the support or between the various active phases in multi-component systems.
6. To identify the binding states of adsorbed species. It is known that some of the adsorbates give rise to multiple adsorption states on the surfaces of solids.
7. Surface chemical changes during transient or non-steady state conditions.

Temperature programmed methods are based on monitoring chemical changes as a function of temperature where temperature is altered linearly with respect to time. Experimentally the system under investigation is charged inside a furnace which is connected to a processor which heats the reactor typically at 0.1 to 60°C/min. A thermal conductivity detector monitors the gas composition. Typically for TPR 5% H_2 in Ar or in TPO 5% O_2 in He is employed to optimize the thermal conductivity difference. Typical schematic representation of the apparatus for TPX measurement is shown in Fig. 3.1.

Table 3.1 Listing of the various thermal analysis techniques and the methods of following the changes in chemical systems

Acronym	Name of the technique	Method
DTA	Differential Thermal Analysis	Temperature difference between the sample and an inert reference like alumina
DSC	Differential Scanning Calorimetry	
TG	Thermo-gravimetry	Weight of the sample
DTG	Differential Thermogravimetry	Weight change with temperature
TMA	Thermomechanical Analysis Dilatometry	Specific volume of the solid sample
TMA	Thermomagnetic Analysis	Magnetic properties
DMC	Differential Microcalorimetry	Enthalpy difference between sample and reference
TPX	Temperature Programmed (X= $\underline{R}$eduction, $\underline{O}$xidation, $\underline{D}$esorption, Sulphidation, or $\underline{S}$urface $\underline{R}$eaction.	Analysis of the gas at the outlet of the unit.

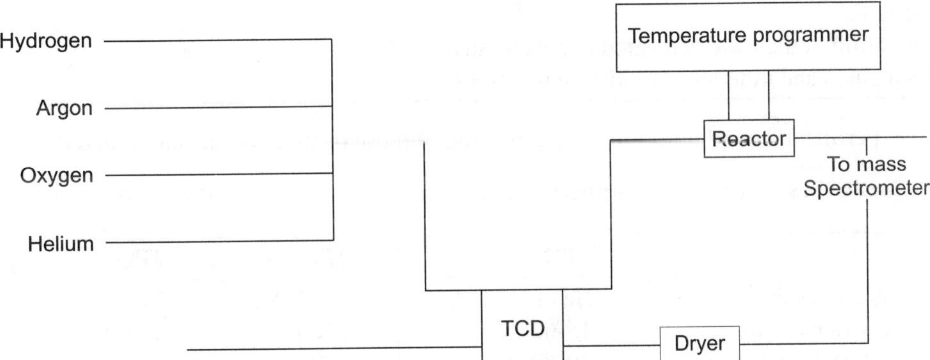

Fig. 3.1 Schematic representation apparatus for TPX measurements

Temperature programmed desorption can be used to identify

1. The nature and strength of adsorption.
2. The various states of adsorption.
3. The quantity of adsorption in a particular adsorption state.
4. The kinetics of adsorption process, like the order of the reaction, the activation energy for the process.
5. In some favourable cases, it is possible to evaluate the reaction network.

The types of information that can be obtained by each of the temperature programmed techniques are given in Table 3.2.

Table 3.2 The types of information that can be generated from various TPX techniques

<table>
<tr><td colspan="1">Temperature Programmed Desorption (TPD)</td></tr>
<tr><td>
1. Identification of the adsorptive properties of materials
2. Characterization of surface functionalities like acidity and basicity
3. Temperature interval where desorption takes place
4. The amount desorbed, adsorption capacity, metal surface determination, dispersion
5. Binding state of the molecules, energetics of adsorption, surface non uniformity
6. Mechanistic and kinetic aspects of adsorption and desorption
</td></tr>
<tr><td>Temperature Programmed Reduction (TPR)</td></tr>
<tr><td>
1. Identification of the redox properties of solids
2. Temperature range of the reduction of the solid
3. Total quantity of the reducing agent consumed, valence state of metal ions in oxidic systems
4. Interaction between metal and the support
5. Indication of the interaction between metals, alloy formation
6. Mechanism and kinetics of reduction process
</td></tr>
<tr><td>Temperature Programmed Oxidation (TPO)</td></tr>
<tr><td>
1. Characterization of the redox properties of metals and metal oxides
2. Characterization of coke species in deactivated catalysts as well as in hydrocarbon conversion reactions
3. Estimation of the coke deposited in deactivated catalysts
4. Mechanism and kinetics of oxidation reactions
</td></tr>
</table>

Typical experimental conditions employed in some of these techniques are summarized in Table 3.3.

Table 3.3 Typical experimental conditions employed in TPX experiments

	TPD	TPR	TPO
Gas composition	He or N_2 or Ar	$5\%H_2/N_2$	$5\%O_2/N_2$
Flow rate ml/min	15-60	15-30	30-90
Heating Rate K/min	10-60	4-60	10-60
Detecting system	TCD or MS	TCD	TCD or MS

3.1 SURFACE ACIDITY DETERMINATION

It is known that many solid acids are better catalysts than even conventional mineral acids and their acid strengths are higher than that of conventional acids. The acidities of these solid catalysts have been evaluated by a number of ways like, titration with a base, [1] (this method has limitations with respect to wettability, phobicity of the solids for the non-aqueous medium used and above all for coloured solids identifying the colour change of the indicator [2]), adsorption of basic probe molecules like ammonia, pyridine, quinolines, use of catalytic reactions like dehydration of alcohols like cyclohexanol as well as by a number of spectroscopic methods like IR and NMR.

TPD is one of the techniques used for evaluating the acidity of solid catalyst. The most often used is the ammonia TPD, since it is easy to perform. However, ammonia adsorption may not differentiate between Bronsted and Lewis acidic sites. Strong interaction of ammonia with acidic site might result in a partial re-adsorption of NH_3, resulting in a shift of the desorption maximum of up to 150°C. Typical adsorption of ammonia on a variety of zeolites is shown in Fig. 3.2.

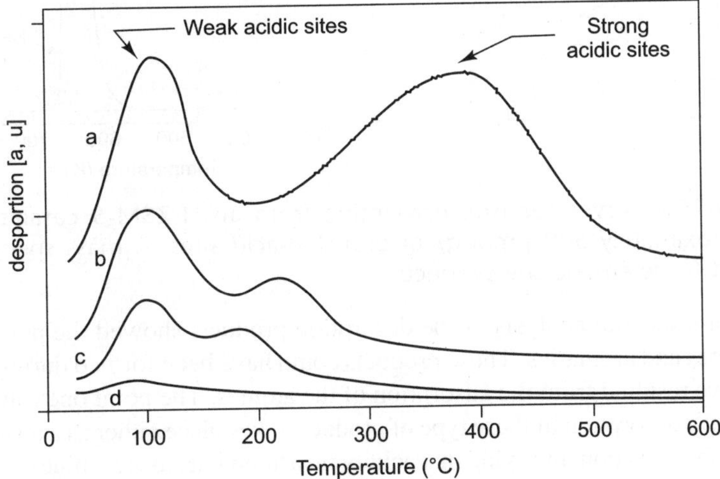

Fig. 3.2 Typical desorption trace for the adsorption of ammonia on various zeolites and other zeolite type substances, (a) Mordenite, (b) ZXM-5, (c) SAPO-11, (d) ALPO-11

The next set of probe molecules are alkyl amines other than methyl amine. They (alkyl amines other than methyl amine) can determine Bronsted acid sites. By Hoffman elimination they also yield olefins and ammonia as products thus the results have to be carefully interpreted.

$$HRNH_2 + H\text{-}O\text{-}Zeol \rightarrow HRNH_3+...\text{-}O\text{-}zeol \rightarrow R+ NH_3 + H\text{-}O\text{-}Zeol$$

The thermal desorption spectra obtained for desorption of n-propyl amine and n-butyl amine are shown in Fig. 3.3. In this figure the X-axis can sometimes be represented in time units instead of temperature and in temperature programmed techniques, time and temperature are related to each other through the heating rate, that is $T = [T_0 + \beta t)]$ where T_0 is the starting temperature, β is the heating rate in degrees per minute and 't' is time in minutes.

It is seen that the desorption spectra for n-propyl amine and n-butyl amine shown in Fig. 3.3 is similar in shape with two identifiable adsorption states, in both cases, marked α and β, however there are distinctive differences. It is seen that in the case of n-propyl amine, the low temperature desorption peak is of equal magnitude as that of the high temperature desorption peak probably indicating that there are two different types of acid sites with different strengths, but the number of sites in each category is almost the same where as in the case of n-butyl amine the number of weak acids appear to be quite small while the total number of strong acid sites is quite high. However, the total number of acid sites available on the surface appears to be the same. The points that emerge from this example are:

1. HY zeolite does possess acid sites that can be probed by organic amines.
2. There are different kinds of acid sites with different strengths.
3. The probe molecules depending on their basicities do not discriminate exactly the different acid strengths.

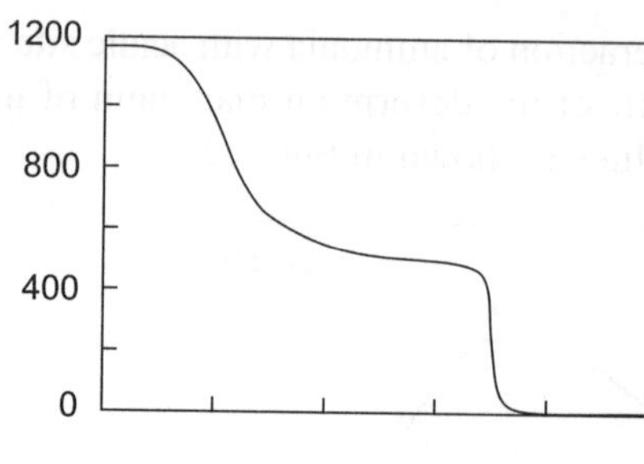
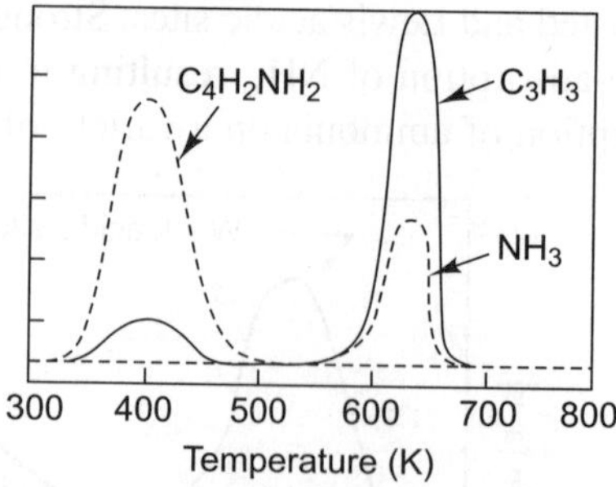

Fig. 3.3 TPD-TGA curves for isopropylamine from an H-ZSM-5 catalyst containing approximately 500 µmol/g/ of Bronsted-acid sites. A mass spectrometer was used to determine the products.

Study of the mass spectral analysis of the desorption products showed the presence of butanes, higher olefins, dienes and aromatics. These products could have been formed during the temperature sweep or could have resulted from the adsorption of the amines. The point one can derive from this observation is that the desorption of these type of products takes place either as a result of temperature sweep or due to surface reaction, implying that acidity measurements using amines as probe molecules has to be treated with caution. However the data given in Table 3.4 show that there is consistency in the surface acidity of HZSM-5 determined by ammonia/ethylamine adsorption. Therefore, there is still hope to rely on acidity determination from the measurement of adsorption of basic probe molecules.

Table 3.4 Comparison of acidity data for HZSM-5 measured by different methods

Sample SiO$_2$/Al$_2$O$_3$ ratio	Carrier gas	Adsorbate	Adsorption Temp. (K)	Acid amount		Acid strength		Ref.
				Mole/g	Sites X 10^{20} per g	Peak 1	Peak 2	
53	N$_2$	NH$_3$	423	0.7304	4.4	476	698	1
				0.6928	4.17	475	649	1
				0.7016	4.27	474	689	1
48	He	C$_2$H$_5$NH$_2$	423	0.7248	4.38	498	698	2
52	He	NH$_3$	423	0.7706	4.64	503	708	3
47.3	He	NH$_3$	423	0.7505	4.52	467	678	4
49.5	He	NH$_3$	343	0.7803	4.70	473	753	5

1. D.C. Zhang and Z.J. Da, Thermochim. Acta, 233 (1994) 87.
2. J.X. Liu, Q.X. Wang, and L.X. Yang, J. Catal., (Chi Hua Xue Bao) 8 (1987) 203.
3. S.Y. Li, G.Y. Gai, Q.W. Wang, X.B. Zhang and G.Q. Che, J. Catal., (Chi Hua Xue Bao) 8 (1987) 47.
4. D.W. Zhov, Q.X. Cao and Q.Z. Li, J. Fuel Chem.Technology, 11(1983) 64.
5. J.G. Post and J.H.C. Van Hooff, Zeolites, 4 (1984) 9.

Table 3.5 Some of the probe molecules that have been used in conjunction with TPD for surface acidity/ basicity determination

Probe molecule	Type of sites normally probed	Some remarks on suitability
Ammonia	The type (Bronsted and Lewis) and concentration of acid sites	Molecular size of ammonia is small, so micro-porous substances can be probed.
Aliphatic amines	The type of acid sites and their concentration	Intra molecular reactions are possible, various products can be identified by TPX techniques.
Pyridine	The type of acid sites and their concentration	Due to size restrictions only some of the pores are accessible. 10 member or higher channels alone are accessible. Availability of molar absorption coefficient value enables quantitative measurements.
Substituted Pyridines	The type of acid sites and their concentration	Accessibility of the sites is restricted; sites near pore openings can be probed.
Benzene, Toluene and Xylenes	Acid strength and accessibility of sites	These molecules can undergo reactions especially isomerization reactions on strong acid sites, Some times slow uptake of the bulky molecules.
Alkanes	Acid strength and accessibility of sites	They may be used to probe sites where hydrocarbon conversion reactions preferentially takes place.
Pyrrole	The type of basic sites and their concentration	Intra molecular reactions can take place hence interpretation has to be careful.
Carbon-dioxide	The type of basic sites and their concentration	Formation of carbonates can vitiate the conclusions.
Nitriles	The type of acid sites and their concentration	Surface reactions can occur and hence interpretation is difficult.
Carbon monoxide	Metal centres and the type of acid sites and their concentration	Low temperatures are necessary; multiple bonded species may also be present.
NO and NO_2	Metal centres (especially ionic sites) and the type of acid sites and their concentration	Nitrate formation on the surface can vitiate the results obtained.

3.2 PREPARATION OF CATALYSTS – APPLICATION OF TEMPERATURE PROGRAMMED METHODS

Preparation of catalysts involves many unit operations like (i) precipitation, (ii) decomposition, (iii) reduction and (iv) impregnation.

In addition, these unit operations are carried out under various operating conditions, like temperature, pH, nature of precipitating agent, nature of the gaseous atmosphere, reducing gas used and its partial pressure, flow rate of the gases used for pre-treatment, moisture content and duration of pre-treatment, nature of support, promoters, impurities present. These will all have profound effect on the nature and texture of the materials that are produced as the final catalytic phase. The effect of some of these parameters will be considered from the point of view of how these techniques provide information on these effects. For example TPR can provide under favourable circumstances, information on the stepwise reduction of the metal ion precursors employed for the preparation of metallic and multi-metallic catalysts. The reduction behaviour of metals can also be elucidated by TPR. The binding of the metal ions with the support if it were to be a variable parameter can be elucidated by TPR. The effect of added promoters and impurities can also be examined from the shift in the position of the reduction signals or appearances of new signals in the TPR traces.

The various applications of TPR technique have been compiled by Hurst et al. [3]. Jenkins and co-workers [4] have examined Pt/SiO_2 catalysts using TPR technique. They have found that the catalyst prepared by ion exchange method gave better dispersion of Pt as compared to the catalyst prepared by impregnation with chloro-platinic acid. The TPR pattern of the two catalysts are shown in Fig. 3.4, wherein it is noticed that the catalyst prepared by ion exchange gets reduced at a higher temperature, higher even than unsupported system and exhibits some structure. It is presumed that the Platinum species in this catalyst is interacting with the silica support. The impregnated catalyst reduced in three widely different temperature regions. The low temperature peak corresponds to was what found for the unsupported system. It is therefore deduced that the poor dispersion of the metal in the impregnated catalyst. The impregnated catalyst was made using chloroplatinic acid, while the ion exchanged catalyst was made using $[Pt(NH_3)_4]$ $(OH)_2$. They have also shown the importance of the calcination step in the case of impregnated catalyst so that the interaction with the support can be established and better dispersion can be obtained.

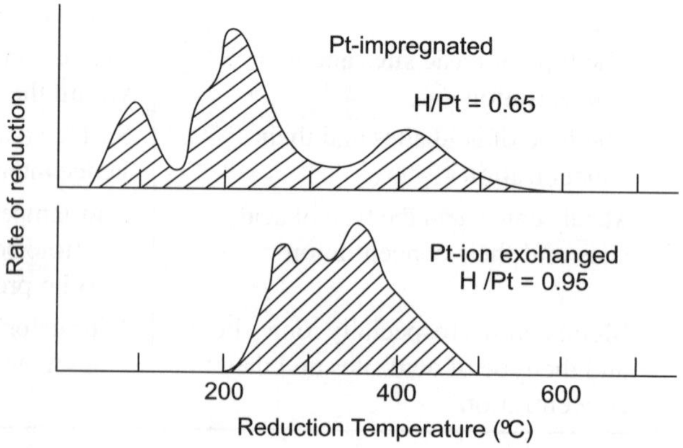

Fig. 3.4 TPR profiles of 0.5 weight % of Pt on SiO_2

It is also possible to use TPR to establish the effect of temperature of calcination on the reduction of metallic species. A typical example is shown in Fig. 3.5. It is seen from this figure that as the calcination temperature is increased, the temperature of reduction also increases showing that the reduction of the Pt species becomes difficult as the calcination temperature increases. Secondly the amount of the reduced Pt species is also changing showing that the extent of Pt species that is freely available for reduction by the gaseous hydrogen is also changing and reducing.

TPR technique can provide information of the type of interaction that can occur in multi-component catalytic systems. Consider the example of bimetallic systems wherein the alloy formation is an important aspect. The alloy formation can occur in any of the four unit operations like impregnation, drying, calcination or reduction stages of the catalyst preparation. In Fig. 3.6, the TPR traces of the Pt-Pd system is shown, wherein one can easily see that the TPR traces of the physical mixture is different from that of the impregnated system. It is seen that Pt associated with Pd also gets reduced at lower temperatures wherein essentially reduction of the palladium predominates. This part of the Pt perhaps associated with Pd in the form of the alloy is also reduced as easily as Pd.

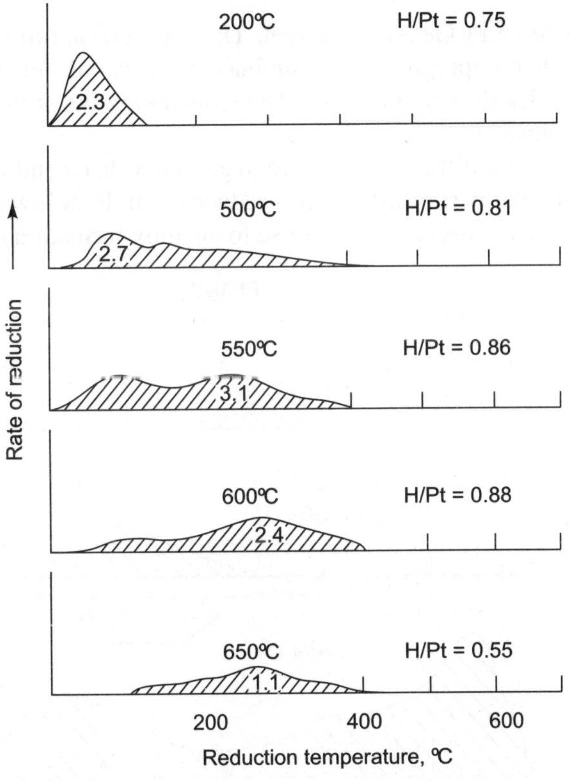

Fig. 3.5 The influence of calcination temperature on Pt/Al_2O_3 catalysts

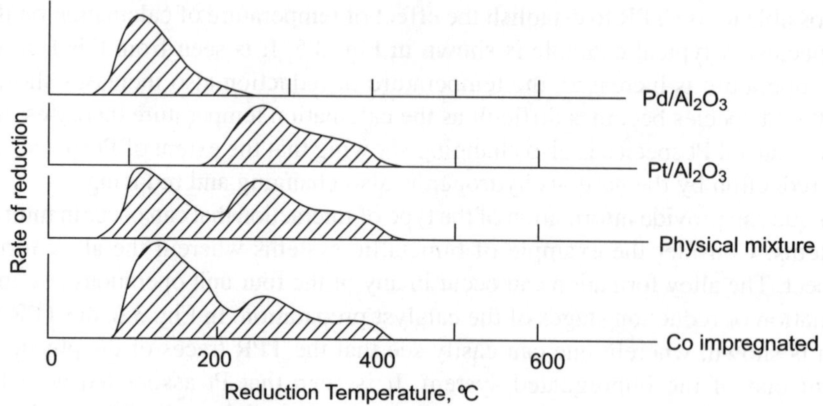

Fig. 3.6 Temperature programmed reduction profiles of various systems evidence for alloy formation in supported catalysts

Another example would be Pt-Ge/Al$_2$O$_3$ system. The data relevant to this system are shown in Fig. 3.7. The Pt and Ge when impregnated on alumina exist as two different species on the support as seen from the TPR profiles shown in Fig. 3.7. The reduction peaks of these two metallic ions are distinct and easily discriminated.

However, after the reduction if this system were to get re-oxidized and reduced, the TPR profile shown in Fig. 3.8 indicate that part of the Ge gets reduced with Pt at low temperatures, indicating these two metallic ions on reduction could give rise to an alloy formation on the surface.

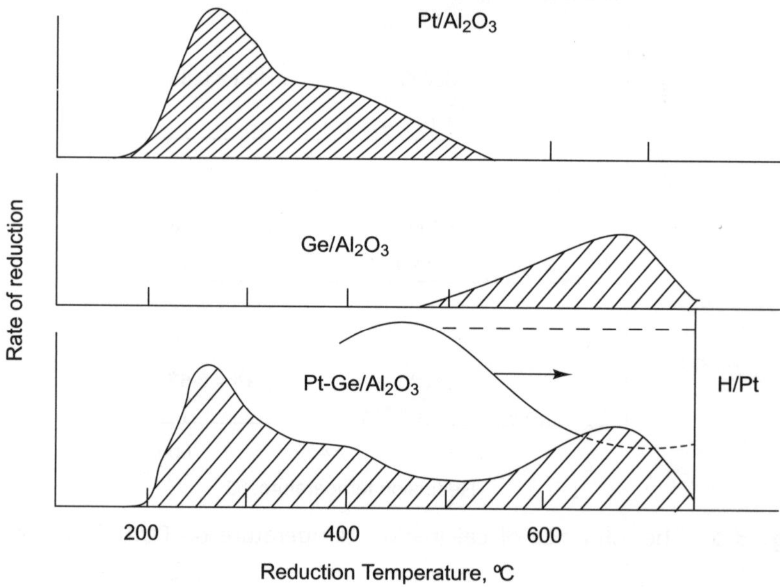

Fig. 3.7 Temperature programmed reduction profile of Pt-Ge/Al$_2$O$_3$ catalyst (Fresh impregnated catalyst)

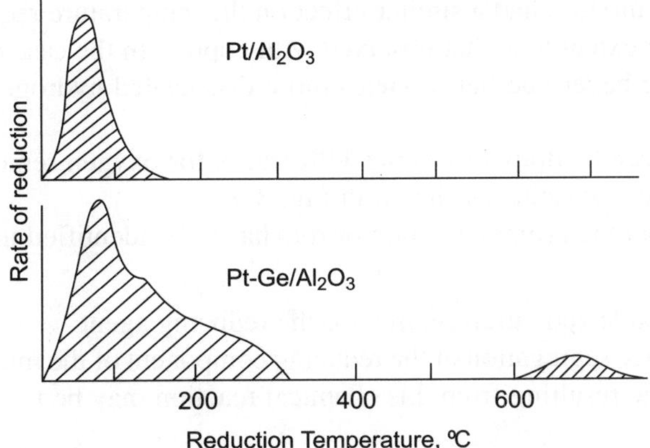

Fig. 3.8 Temperature programmed reduction profile of Pt/Al$_2$O$_3$ and Pt-Ge/Al$_2$O$_3$ system after re-oxidation (Alloy formation during reduction)

Senzi Li et al. [5] have examined promoted iron catalysts for Fischer Tropsch synthesis. They have compared the reduction and carbonization behaviour of Fe-Zn-K oxides promoted by Ru or Cu. The oxygen removal rates of these two promoted systems (TPR traces) with that of the impregnated systems are shown in Fig. 3.9.

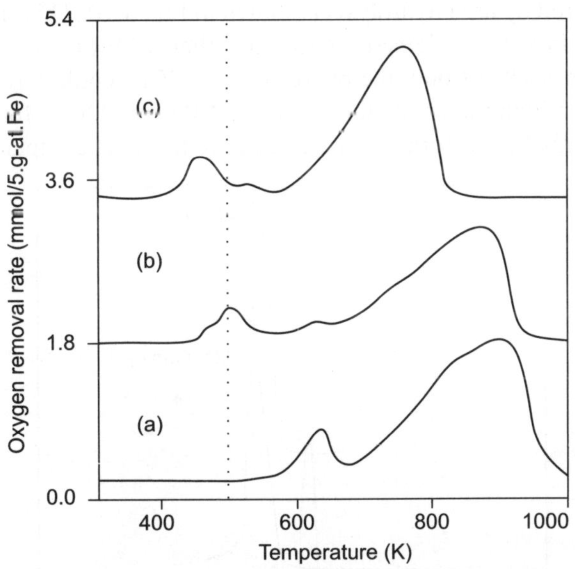

Fig. 3.9 Oxygen removal rates for Fe–Zn–K–Cu(Ru) oxides in H$_2$. (a) Fe–Zn–K$_2$, (b) Fe–Zn–K$_2$–Cu1, and (c) Fe–Zn–K$_2$–Ru1. (0.2 g sample; K/Fe=0.02, Cu(Ru) /Fe = 0.01; 0.167 K/s ramping rate; 20% H$_2$/Ar, 0.268 mol/h flow rate)

The reduction sequence that is Fe$_2$O$_3$ → Fe$_3$O$_4$ → FeO → Fe is followed in all the three catalytic systems. The addition of Cu to Fe-Zn-K system reduced the temperature of reduction of Fe$_2$O$_3$ by 130 K from around 620 K to nearly 490 K; this reduction in temperature has been attributed to the formation of hydrogen dissociation sites generated by the reduction of CuO to Cu metal at low temperatures. Ru in the mixture had a similar effect on the temperature required for this reduction of Fe_2O_3 but to a greater extent than that observed with copper. In the case of ruthenium, the metal dispersion appears to be better and hence yields more dissociated hydrogen for easy reduction of Fe_2O_3.

Similar conclusions can be drawn from the TPR traces for oxygen removal by CO and carbon introduction rates on these systems as shown in Fig. 3.9.

It must be realized that the promoter action or role has to be identified in many ways. They can be listed as follows:

1. Generation of suitable activation centres for the reducing agent.
2. They can also induce segregation of the reducing component of the multi-component systems.
3. The entropy change resulting from this chemical reaction may be favourable in the presence of the promoter.
4. The promoter species could also function as effective heat transferring medium or sink or source so as to facilitate the facile reduction of the precursor species.

3.3 STUDIES ON THE PHASE COMPOSITION – TPR STUDIES

TPR technique can also be utilized to distinguish the reduction behaviour of two phases of the same substance. For example NiO can crystallize in cubic as well as rhombohedral forms. The TPR traces of Ni (OH)$_2$ prepared by urea hydrolysis is shown in Fig. 3.10. This shows a simple reduction peak. But if the same hydroxide is calcined at 673 K and then a TPR trace is recorded one can easily distinguish the reduction peaks of both the phases of NiO. This could have arisen since while the hydroxide is decomposed both the crystalline forms of NiO could have formed. Another aspect to be noted is that the rhombohedral form is easier to get reduced as compared to the cubic form.

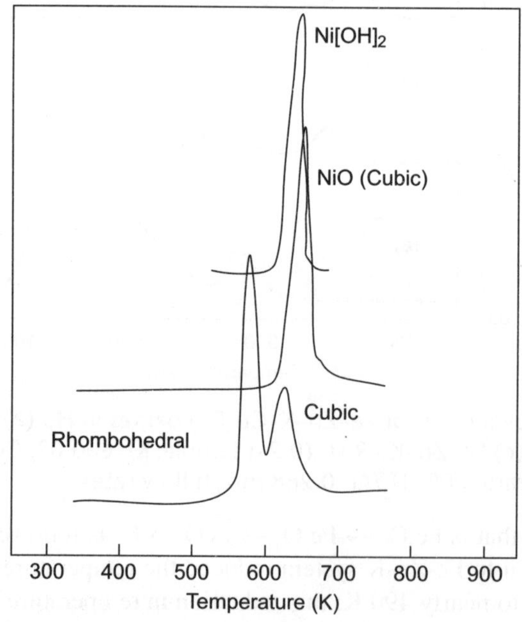

Fig. 3.10 Temperature programmed reduction profiles of Ni(OH)$_2$ and of NiO

The TPR traces for the hydroxide of nickel prepared by different impregnation methods are shown in Fig. 3.10. There are distinguishable nickel reduction stages that can be identified in these traces. They are possibly free NiO in rhombohedral and cubic forms, NiO that is reduced by an autocatalytic process and a bound state of NiO which gets reduced at higher temperatures. Thus one can see finer details of interaction of the metallic species with the support from the TPR traces. TPR studies of zeolite systems especially metal exchanged systems have been interesting since it could identify the exact site wherein the metal ions are exchanged, how the metal ions migrate over the zeolite surface on re-oxidation and thus it provides information on sub surface mobility and diffusion. [8]

Examples of the application of the Temperature Programmed Reaction Spectroscopy (TPRS) will be considered.

The decomposition of deteurated formic acid on Ni (110) surface has been investigated by Madix [9]. The desorption pattern obtained is shown in Fig. 3.11. One can see there are two distinct CO desorption peaks, indicating the desorption of CO is taking place from two reaction steps probably one coming out of the direct decomposition of the adsorbed formic acid molecule and the high temperature one from the adsorbed CO.

The overall reaction sequence as postulated is

$$2DCOOH \rightarrow [DCO + DCOO] + H_2O$$

$$DCO \rightarrow CO + D$$

$$DCOO \rightarrow CO_2 + D$$

$$[CO + DCOO]_{ads} \rightarrow CO_{ads} + CO_2 + D$$

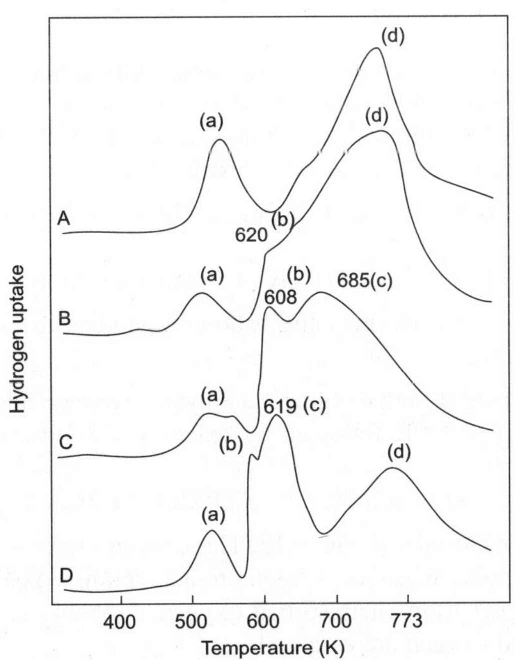

Fig. 3.11 Temperature programmed profiles of Ni/Al$_2$O$_3$ catalyst precursors (A) EAL-673; (B) EAE-673; (C) EAA-673; (D) EEQ-673; Reducing species (a) NiO (Rh); (b) NiO (Cubic); (c) NiO (Auto catalysis); (d) NiO (bound)

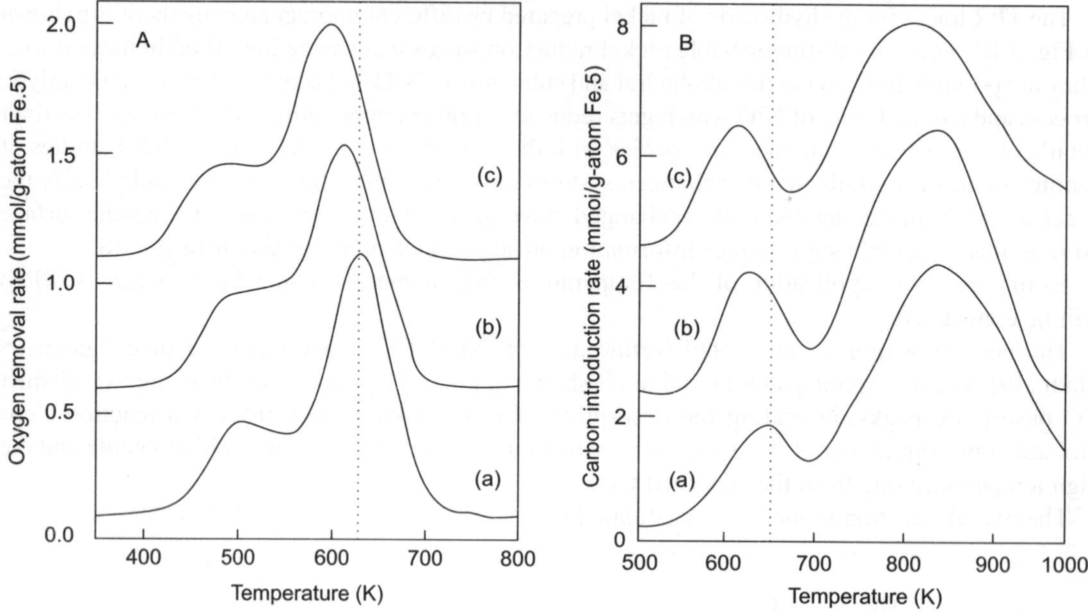

Fig. 3.12 (A) Oxygen removal and (B) carbon introduction rates for Fe–Zn–K–Cu(Ru) samples in CO. (a) Fe–Zn–K2, (b) Fe–Zn–K2–Cu1, and (c) Fe–Zn–K2–Ru1. (0.05 g sample; K/Fe=0.02, Cu(Ru)/Fe = 0.01; 0.167 K/s ramping rate; 20% CO/Ar, 0.268 mol/h flow rate)

This type of detailed reaction steps could be elucidated from the analysis of the products evolving from the surface as a function of temperature programming.

One of the problems which requires information is the reaction of oxygen with ethylene for the formation of ethylene oxide. The questions to be answered are:

1. Which oxygen species is involved in the formation of the desired product namely the ethylene oxide?
2. How does the adsorbed oxygen species involved in the reaction?

Van Santen et al. [10] have performed the following studies to establish these points. The observations made by them are:

1. No reaction was observed if both oxygen and ethylene were co-adsorbed at 113K.
2. Adsorption of oxygen at 300K followed by reaction with ethylene produced only carbon dioxide.
3. Adsorption of oxygen at 470K followed by a TPRS with ethylene produced ethylene oxide.

These results are shown pictorially in Fig. 3.12. These studies have revealed that ethylene oxide formation has been possible only when the oxygen pre-adsorption temperatures are high enough to form subsurface oxygen apart from chemisorbed oxygen. Probably for selective oxidation both subsurface and chemisorbed oxygen are essential.

Another example is the reaction between CO and hydrogen on potassium promoted ruthenium catalysts studied using TPRS technique using IR as the detection mode. The TPRS trace obtained by Gonzalez and Miura [11] is shown in Fig. 3.13. It can be seen that the CO which shows absorption band around 2020 cm^{-1} is not reactive to produce methane while the CO showing IR absorption band around 1950 cm^{-1} is reactive and form methane. From these studies it may be concluded that the linearly adsorbed CO is not participating in the reaction while the CO adsorbed in the bridged mode is reactive for the hydrogenation reaction.

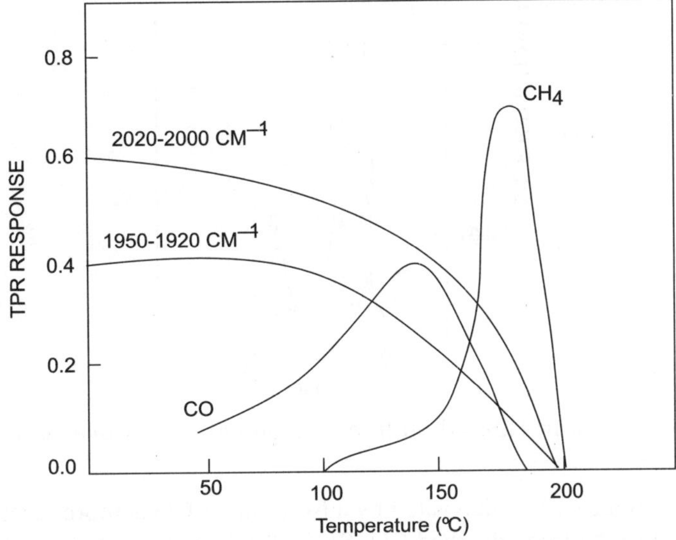

Fig. 3.13 Simultaneous IR-TPRS for reaction CO$_{ads}$ + H$_2$ on Ru-K Catalyst

The carbon deposition in hydrocarbon reactions is an important step and various attempts have been made to identify the nature of carbon that is formed on the catalyst surface. The results obtained on the surface of Ni/Al$_2$O$_3$ during the hydrogenation of CO are shown in Fig. 3.14.

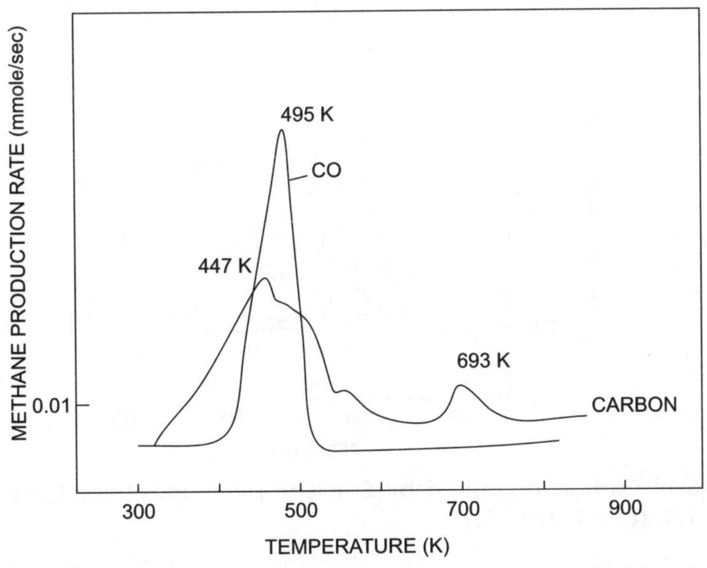

Fig. 3.14 TPRS of adsorbed CO (300K) and surface carbon deposited by CO

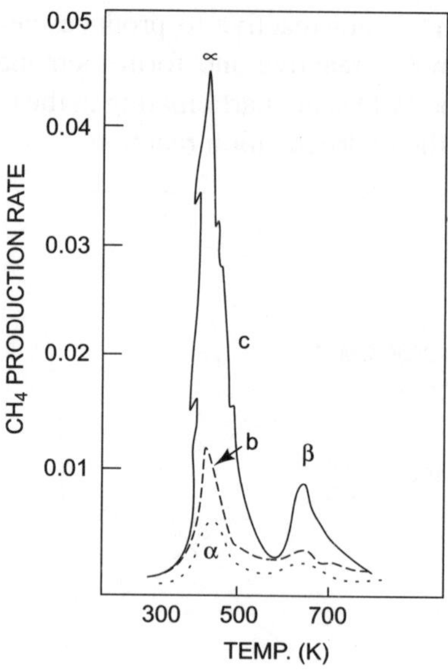

Fig. 3.15 TPRS of C_{ads} with H2-Carbon from CO on Ni/Al$_2$O$_3$; Coke % (a) 0.48, (b) 1.19, (c) 3.14%

It is seen that the surface carbon deposited by adsorption of CO is more reactive towards hydrogenation as compared to the free adsorbed CO. Secondly the nature of the carbon deposited from the adsorption of CO is different from the carbon deposited by the adsorption of ethylene in terms of reactivity towards hydrogenation (See Fig. 3.15).

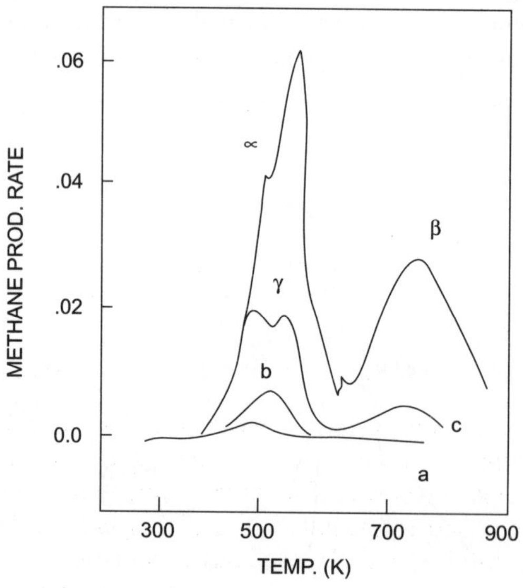

Fig. 3.16 TPRS of Carbon deposited by Ethylene decomposition; Coke : % (a) 0.45, (b) 1.4, (c) 3.7, (d) 11.1

In fact more than one form of carbon has been identified by the adsorption of ethylene and their reactivity towards hydrogenation also differ. (Refer to Fig. 3.16.)

The formation and nature of coke formed can also be probed by temperature programmed oxidation. Parera et al. [12] have investigated the combustion of carbon deposited on Pt/Al_2O_3 catalysts. Their results are shown in Fig. 3.17. It is seen that the coke on the metal (Pt) is oxidized at a lower temperature (553K) while the coke on the support is oxidized only at 763K. Another observation worth noting is that the temperature of combustion of the coke decreases with increase in the metal loading. Similar observations have been recorded by others as well [13, 14].

Many of the partial oxidation catalysts like molybdates, vanadates, tungstates and antimonates are amenable for easy reduction and oxidation, and the ease of reduction is a measure of their activity and the re-oxidation capability is a measure of the life of the catalyst. Miura et al [15] have studied the TPRO of Bi_2O_3: MoO_3 and bismuth molybdate catalyst. The results obtained by them are shown in Figs. 3.18 and 3.19.

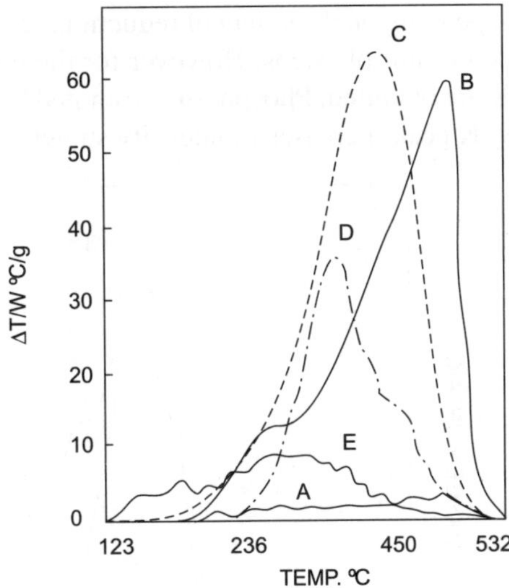

Fig. 3.17 TPO of the coked catalysts containing different loadings of Pt-Al_2O_3 (a) 0.0%, (b) 0.37%, (c) 1.6%, (d) 0.75%, (e) only Pt

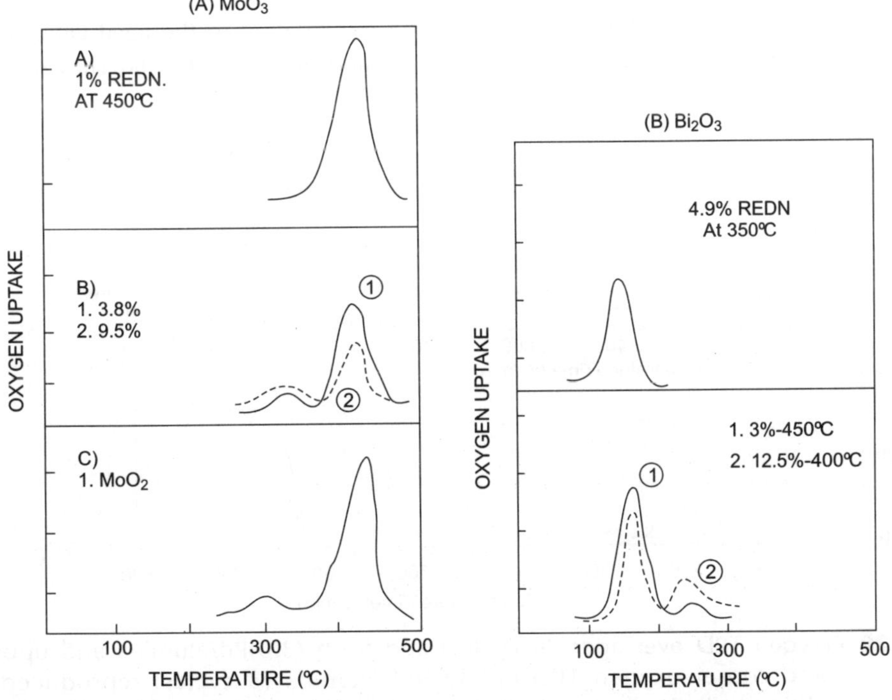

Fig. 3.18 Temperature Programmed Reoxidation of (a) MoO_3 and (b) Bi_2O_3

Depending on the extent of reduction, re-oxidation of Bi and Mo takes place at 443K and 693K in single or multiple steps. However for the bismuth molybdate, two re-oxidation peaks at 593K and 673K are obtained. Phosphorus promoted bismuth molybdate however shows the re-oxidation peak at 593K peak increases in intensity showing phosphorus addition facilitates the re-oxidation.

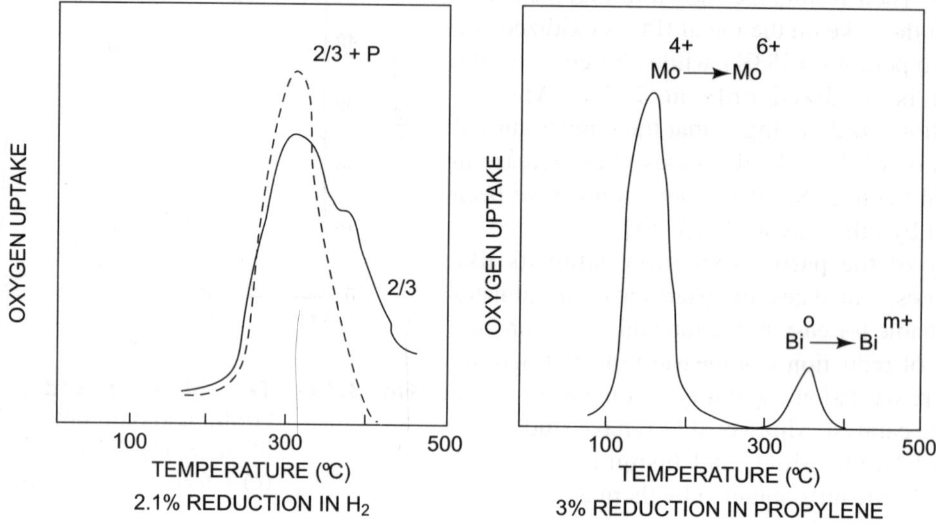

Fig. 3.19 TPRO of Bismuth molybdates

Another example is Pd on alumina catalysts. The re-oxidation of the aged catalysts showed redistribution of Pd particles on the used catalyst. The results of such a study are shown in Fig. 3.20.

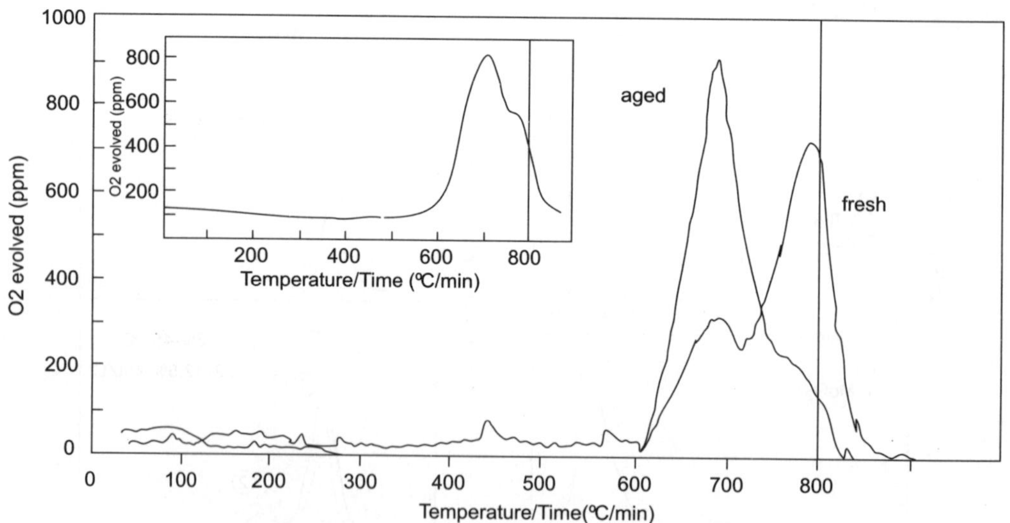

Fig. 3.20 Oxygen TPD over aged (4 cycles) and fresh (3.0)Pd/alumina and upper left hand corner oxygen TPD of (3.0)Pd/alumina (3 cycles) reproduced from F. Klingstedt, H. Karhu, A. Kalantar Neyestanaki, L.E Lindfors, T. Salmi and J. Vayrynen, J. Catalysis, 206,(2002)248

REFERENCES

1. H.A. Benesi, J. Am. Chem. Soc., *78*(21) (1956) 5490 - 5494
2. Chen Hang Sun and John C. Berg, Advances in Colloid and Interface Science, 105 (2003) 151
   M. Frenkel, Analytical Chemistry, 47 (1975) 598
3. Nicholas W. Hurst, Stephen J. Gentry , Alan Jones , Brian D. McNicol, Catalysis Rev. Sci and Engg., 24 (1982) 233
4. J.W. Jenkins, B.D. McNicol, and S.D. Roberson, Chem. Technol., 7 (1977) 316
5. Li S., Krishnamoorthy S., Li A., Meitzner G.D., Iglesia E.J. Catal, 206 (2) (2002) 202
6. J.A.Z. Pieterse, R.W. van den Brink, S. Booneveld and F. A. de Bruijn, Applied Catalysis, B46 (2003) 239
7. Xiang Wang, Haiying Chen and W.M.H. Sachtler, Applied Catalysis, B29 (2001) 47; B26 (2000) L227
8. K.J. Leary, J.N. Michaels and A.M. Stacy, J. Catalysis, 107 (1987) 393
9. Madix, R.J., Catal. Rev. Sci. & Engg. 26 (1984) 281
10. C. Backx, J. Moolhysen, P. Geenen and R.A. Van Santen, J. Catal., 72 (1981) 364
11. R.D. Gonzalez, and H. Miura, J. Catal., 77 (1982)338
12. J.M. Parera, N.S. Figoli, and E.M. Traffano, J. Catal., 79 (1983) 481
13. J. Barbier, P. Marecot, N. Martin, N. Elassel, and R. Maurel, in Catalyst Deactivation B. Delman and G.F. Firnebt, Eds., Elseveir, Amsterdam (1980) 53
14. R.L. Mieville, J. Catal., 100 (1986) 482
15. H. Miura, Y. Morikawa, and T. Sheiasaki J. Catal., 39 (1975) 22

Introduction to Electron Spectroscopy

Photoemission from solids is one of the versatile techniques that can be advantageously used for characterization of a number of properties of the solid substrate especially catalysts and their surfaces. The phenomenon of photoemission is one of the direct proofs of the Einstien's postulate. Three parallel endeavours led to the development of photoemission as a technique for solid state chemistry. They are:

1. Siegbahn and his group developed improved energy resolution of the electron spectrometers and combined it with X-rays.
2. Turner and his co-workers applied photoelectric effect to gases and by using ultraviolet line from helium source they could resolve vibrational fine structure of the electronic levels.
3. Spicer measured photoelectron spectra from solids in vacuum irradiated with UV light but the LiF window used cut off all the photons with energies greater than 11.6 eV and hence he could only get a small part of the valence band. Eastman and Cashion extended the energy range to about 40 eV by differential pumping and windowless UV sources.

The availability of synchrotron sources has extended the photon energy range further such that all energies between the UPS and XPS regime are available. The mean free path of the electrons decides the depth to which the material can be probed. Fig. 4.1 shows the mean free path λ, of electrons in elemental solids and it is found to depend on the kinetic energy of the electrons and it is of the order of 1-2 nm for kinetic energies in the range 15-1000 eV and this is an ideal situation for surface sensitivity.

XPS and UPS are two techniques based on photoelectric effect. When an atom absorbs a photon of energy $h\nu$, then a core or valence electron with binding energy E_b is ejected with the Kinetic energy

$$E_k = h\nu - E_b - \varphi$$

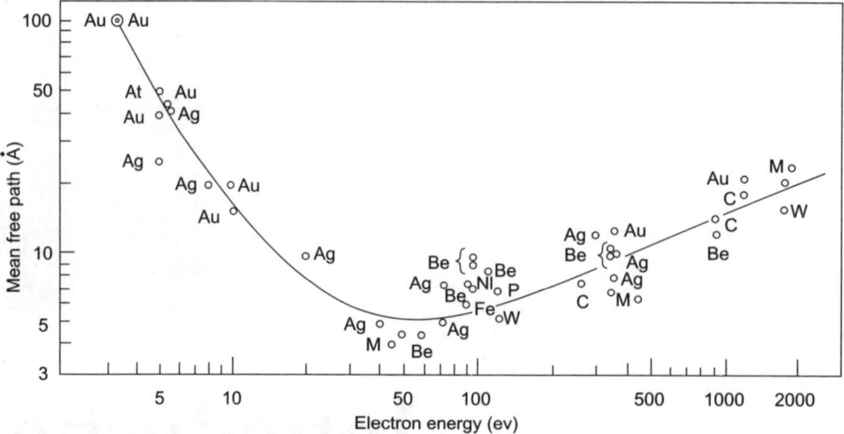

Fig. 4.1 Universal curve for electron mean free path as a function of the electron energy. The ability of electrons to penetrate into and/or escape from materials is strongly energy dependent. At energies of » 50 eV the mean free path reaches a minimum

where E_k is the Kinetic energy of the ejected electron, h is the Planck's constant, ν is the frequency of the exciting radiation, E_b is the binding energy of the photoelectron with respect to the Fermi level of the sample and f is the work function of the spectrometer. In XPS spectrum one measures the intensity of photoelectrons N (E) as a function of their kinetic energy or more often in terms of binding energy. The photoelectron peak designations are based on the quantum numbers of the levels from which the photoelectron is originating. Spin orbit splitting as well as binding energies of a particular electron energy level increase with increasing atomic number. The nomenclature used is shown in Table 4.1. The intensity ratio of the spin orbit split doublet peaks is determined by the multiplicity of the corresponding levels and is equal to $2j \pm 1$. For example the $3d_{5/2}$ and $3d_{3/2}$ peaks will have an intensity ratio of 6:4 where as the $4f_{7/2}$ and $4f_{5/2}$ peaks will have an intensity ratio of 8:6 and $2p_{3/2}$ and $2p_{1/2}$ will have intensity ratio 4:2.

Table 4.1 Spectroscopic Notations used in XPS and also in AES

n	L	j	X-ray level	Electron level
1	0	1/2	K	$1s$
2	0	1/2	L_1	$2s$
2	1	1/2	L_2	$2p_{1/2}$
2	1	3/2	L_3	$2p_{3/2}$
3	0	1/2	M_1	$3s$
3	1	1/2	M_2	$3p_{1/2}$
3	1	3/2	M_3	$3p_{3/2}$
3	2	3/2	M_4	$3d_{3/2}$
3	2	5/2	M_5	$3d_{5/2}$
4	3	5/2	N_6	$4f_{5/2}$
4	3	7/2	N_7	$4f_{7/2}$

4.1 COMPOSITION ANALYSIS

The general expression for the intensity of an XPS peak is

$$I = F_x S(E_k)\, \sigma(E_k) \int n(z) \exp\left[-z/\lambda(E_k)\cos\theta \right] dz$$

where $\sigma(E_k)$ is the cross-section for emission of a photoelectron in the direction of the analyzer from the relevant inner shell per atom of X by a photon of energy $h\nu$. F_x is the flux of the X-ray line per unit area at the sample and $S(E_k)$ is a spectrometer dependent function containing the detection efficiency, $n(z)$ denotes the atomic density as function of the depth in the sample, z and the depth distribution function (DDF) is characterized by the inelastic mean free path (IMFP) $\lambda(E_k)$ of the photoelectrons.

In the case of a homogeneous concentration through out the sample the following expression will hold for the intensity of emission for each element.

$$I = F_x S(E_k)\, \sigma(E_k)\, n\lambda(E_k)\cos\theta$$

Cross section values have been tabulated by Scofield. He has tabulated extensive list of X-ray cross sections for the sub shells of electrons in the elements of the periodic chart and has specficially utilized X-ray energies of 1254 and 1487 eV to produce absorption coefficients relevant to PES measurements. These data have been normalized to a relative atomic sensitivity of 1.00 for carbon. One such compilation is available in ref 1. Mean free path is a function of the material. For example the Si 2p electrons have a mean free path 3.7 nm in SiO_2 and 3.2 nm in Si. Appropriate back ground corrections have to be carried out for determining the intensity of the photoemission peak.

4.2 BINDING ENERGIES AND OXIDATION STATES

Binding energies can be used not only to identify the chemical element but also the oxidation state of the element since the binding energy of core electron depends on the chemical state of the element as well. Chemical shifts are in the range of 0-3 eV. In general the binding energy increases with increase in oxidation state and for a fixed oxidation state with the electronegativity of the ligands like FeF_3, $FeCl_3$, $FeBr_3$ and so on. Typical data collected from literature are given in Tables 4.2 and 4.3.

Table 4.2 Typical observed binding energy for C 1s and oxygen 1s in various organic compounds showing the chemical shifts

Functional group	Structural fragment	Binding energy, eV
Hydrocarbon	C-H, C-C	285.0
Amine	C-N	286.0
Alcohol, ether	C-O-H, C-O-C	286.5
Cl bound to carbon	C-Cl	286.5
F bound to carbon	C-F	287.8
Carbonyl	C=O	288.0
Amide	N-C=O	288.2
Acid, ester	O-C=O	289.0
Carbamate	N-C=O	289.6

(Contd.)

Table 4.2 Typical observed binding energy for C 1s and oxygen 1s in various organic compounds showing the chemical shifts

Functional group	Structural fragment	Binding energy, eV
Carbonate	$\underline{C}O_3^{2-}$	290.3
2F bound to carbon	$-(CH_2-\underline{C}F_2-)$	290.6
3F bound to carbon	$-\underline{C}F_3$	293.2
Carbonyl oxygen	$C=\underline{O}$	532.2
Alcohol, ether oxygen	$C-\underline{O}-H, C-\underline{O}-C$	532.2
Ester	$C-\underline{O}$	533.2

Table 4.3 Typical Binding energy (chemical shift) data for a set of inorganic compounds

Species	Level probed	Binding Energy, eV
Li	1s	55.1
Li_2O	1s	55.6
LiF	1s	55.7
Be	1s	111.6
BeF_2	1s	115.3
BeO	1s	113.7
B	1s	186.5
B_2O_3	1s	193.3
C	1s	284.5
CO_2	1s	291.9
CF_4	1s	296.7
N	1s	398.4
NH_3	1s	398.8
O_2/Mo	1s	530.2
Al_2O_3	1s	531.6
F	1s	685.7
LiF	1s	685
Na	1s	1070.8
NaF	1s	1071.3
Mg	1s	1303.2
MgO	1s	1303.9
Mg	2p	49.6
MgO	2p	50.3
Al	2p	72.8
Al_2O_3	2p	74.6
Si	$2p_{3/2}$	99.5
SiO_2	$2p_{3/2}$	103.3
K	$2p_{3/2}$	294.7
KF	$2p_{3/2}$	293.1
Ca	$2p_{3/2}$	345.9

(Contd.)

Table 4.3   Typical Binding energy (chemical shift) data for a set of inorganic compounds

Species	Level probed	Binding Energy, eV
CaF_2	$2p_{3/2}$	347.9
Ti	$2p_{3/2}$	454.0
TiO_2	$2p_{3/2}$	459.0
V	$2p_{3/2}$	512.4
V_2O_5	$2p_{3/2}$	517.4
Cr	$2p_{3/2}$	574.6
Cr_2O_3	$2p_{3/2}$	576.6
Mn	$2p_{3/2}$	638.8
MnF_3	$2p_{3/2}$	6426
Mn_2O_3	$2p_{3/2}$	641.5
Fe	$2p_{3/2}$	706.8
Fe_2O_3	$2p_{3/2}$	711.0
FeO	$2p_{3/2}$	709.6
Co	$2p_{3/2}$	778.1
CoF_2	$2p_{3/2}$	783
Co_3O_4	$2p_{3/2}$	780.2
Ni	$2p_{3/2}$	852.5
NiO	$2p_{3/2}$	854.4
Cu	$2p_{3/2}$	933.1
CuO	$2p_{3/2}$	934.1
Zn	$2p_{3/2}$	1021.6
ZnO	$2p_{3/2}$	1021.8
Ga	3d	18.5
Ga_2O_3	3d	19.7
Ge	3d	29.4
GeO_2	3d	33.2
As	3d	42.05
As_2O_3	3d	45.8
Ag	3d	374
Ag_2So_4	3d	374.2
Cd	3d	411.5
CdO	3d	411.7
In	3d	451.8
In_2O_3	3d	452.3
Sn	3d	492.9
SnO_2	3d	494.6
Sb	3d	537.3
Sb_2O_3	3d	539.2
Te	3d	583.7
TeO_2	3d	587.2
Mo	$3d_{5/2}$	227.8

(Contd.)

Table 4.3   Typical Binding energy (chemical shift) data for a set of inorganic compounds

Species	Level probed	Binding Energy, eV
MoO_3	$3d_{5/2}$	232.4
Pd	$3d_{5/2}$	335
PdO	$3d_{5/2}$	337
Ta	$4f_{7/2}$	21.7
Ta_2O_5	$4f_{7/2}$	26.0
Pt	$4f_{7/2}$	71.2
PtO	$4f_{7/2}$	74.6

Data complied from http://srdata.nist.gov/xps/Bind_E.asp

4.3 FINAL STATE EFFECTS

Consider an element with N electrons, with a total energy $E_N^{\,1}$ in the initial state. The atom absorbs a photon of energy $h\nu$; the absorption event takes less than 10^{-17} s. Some 10^{-14} s later, the atom has emitted the photoelectron with kinetic energy E_k and is itself in the final state with one electron less and a hole in one of the core levels.

$$E_N^{\,1} + h\nu = E_{n-1,1}^{\,f} + E_k$$

$E_N^{\,1}$ is the initial total energy of the atom with N electrons in the initial state i.e., before the photoemission, $h\nu$ is the energy of the photon,

$E_{n-1,1}^{\,f}$ is the total energy of the atom with N-1 electrons and a hole in the core level 1 in the final state, E_k is the Kinetic energy of the photo ejected electron.

Rearranging one gets

$$E_b = h\nu - E_k - \varphi = E_{n-1,k}^{\,f} - E_N^{\,i} - \varphi$$

The remaining N-1 electrons in the final state as well as the electrons in the neighbouring atoms feel the existence of a core hole. These results in lowering the total energy of the atom by an amount ΔE_{relax}. This relaxation energy has intra and extra atomic contributions and is included in the kinetic energy of the photoelectron. Thus the binding energy is not equal to the orbital from which the photoelectron is emitted; the difference is caused by the reorganization of the remaining electrons when an electron is removed from an inner shell.

Therefore the binding energy has both the information, that is the state of the atom before the photo ionization (**initial state**) and on the core ionized atom left behind, after the emission of an electron (**final state**).

In most cases it is possible to consider the initial state effects alone. The charge potential model explains the physics behind such binding energy shifts by means of the formula

$$E_b^{\,I} = kq_i + \Sigma q_j / r_{ij} + E_b^{\,ref}$$

$E_b^{\,I}$ is the binding energy of an electron from an atom I, q_i is the charge on the atom, K is a constant, q_j is the charge on a neighbouring atom j, r_{ij} is the distance between atom i and atom j, $E_b^{\,ref}$ is the suitable energy reference.

The first term indicates that the binding energy increases with increase of positive charge on the atom from which the photoelectron originates. In ionic solids, the second term counteracts the first term because the charge on a neighbouring atom will have the opposite sign. Because of its similarity to the lattice potential in ionic solids, the second term is referred as the Madelung sum.

4.4 REFERENCE

Normally the sample and the spectrometer are in electrical contact if the sample is conducting. The Fermi levels of these two systems will be the same. If an electron were to be ejected from the substance say a metal from its Fermi level, it is still bound to the metal and these electrons have an ionization potential equal to the work function of the metal. However the electrons coming from the Fermi level of the sample are detected with a Kinetic energy equal to $h\nu - \varphi_{sp}$ where φ_{sp} is the work function of spectrometer and not that of the sample. The removal of an electron from the Fermi level of the sample will require φ the work function of the sample. The liberated electron is now at the potential outside the sample but not yet at the absolute vacuum level. Thus on its way to the analyzer the electron is accelerated or decelerated by the work function difference which is the contact potential between the sample and the spectrometer.

The description given has assumed that the photoemission event occurs sufficiently slowly to ensure that the escaping electron feels the relaxation of the core-ionized atom. This is what is called the adiabatic limit. All the relaxation effects of the energetic ground state of the core-ionized atom are accounted for in the kinetic energy of the photoelectron. In other extreme, the 'Sudden limit', the photoelectron is emitted immediately after the absorption of the photon before the core ionized atom relaxes. This often is accompanied by shake-up, shake-off and plasmon loss processes, which give additional peaks in the spectrum.

Shake-up, shake-off losses are final state effects which arise when the photoelectron imparts energy to another electron of the atom. This electron ends up in a higher unoccupied state (shake-up) or in an unbound state (shake-off). As a consequence, the photoelectron looses kinetic energy and appears at a higher binding energy in the spectrum. Discrete shake-up losses are prominently present in the spectra of oxides and many other compounds. They have diagnostic value as the precise loss structure depends on the environment of the atom. In metals shake-up of electrons in the valence band to empty states at the Fermi level gives a continuous range of energy losses from zero to the energy of the bottom of the band. The result is that the XPS peaks of the metals such as Rh and Pt are asymmetrically broadened towards higher binding energies. As the probability of valence band shake-up depends on the density of states at or just above the Fermi level, the effect is most pronounced in d-metals, but hardly of significance for s-metals such as Cu, Ag and Au which all give sharp symmetric peaks. In fact the narrow Ag 3d lines are often used to demonstrate the energy resolution of the spectrometer.

Multiplet splitting occurs if the initial state atom contains unpaired electrons. Upon photoemission this electron may interact through its spin moment with the spin of the additional unpaired electron in the core level from which the photoelectron left. Parallel and anti-parallel spins give final states, which differs by 1-2 eV in energy. A well known example is the splitting of the N 1s and oxygen 1s peaks normally single lines – in the spectrum of gas phase NO due to an unpaired spin in the 2π level in the NO molecule.

Note that multiple splitting is a final state effect, whereas the spin-orbit splitting of p, d and f levels is an initial state effect.

The line width is determined by the line width of the X-ray source, the broadening due to the analyzer and the natural line width of the level under study. These three factors are related as follows

$$(\Delta E)^2 = (\Delta E_x)^2 + (\Delta E_{an})^2 + (\Delta E_{nat})^2$$

(ΔE) is the width of a photoemission peak at half maximum, (ΔE_x) is the line width of the X-ray source, (ΔE_{an}) is the broadening due to the analyzer, and (ΔE_{nat}) is the natural line width.

ΔE_x, the line width of the X-ray source is of the order of 1 eV for Al or MgK_{α} sources, but can be better than about 0.3 eV with the use of monochromator.

The monochromator narrows the line significantly and focuses it onto the sample. It also cuts out all unwanted X-ray satellites and background radiation. Important advantage of using a monochromator is that heat and secondary electrons generated by the X-ray source cannot reach the sample.

The broadening due to the analyzer depends on the energy at which the electrons travel through the analyzer and the width of the slits between the energy filter and the actual detector. The analyzer contribution to the line width becomes irrelevant at low pass energies however at the cost of intensity.

Heisenberg's uncertainty relation determines the natural line width

$$\Delta E_{nat}. \; \Delta t = h/2\pi$$

The lifetime of the core-ionized atom is measured from the moment it emits a photoelectron until it decays by Auger processes or X-ray fluorescence. As the number of decay possibilities for an ion with a core hole in a deep level (e.g. 3s level) is greater than that for an ion with core hole in a shallow level (e.g. 3d level) a 3s peak is broader than a 3d peak.

4.5 CHARGING AND SAMPLE DAMAGE

Samples that eject photoelectrons give rise to accumulation of charge. The potential the sample acquires is determined by the photoelectric current of electrons leaving the sample, the current through the sample holder towards the sample and the flow of Auger and secondary electrons from the source window onto the sample. Due to the accumulation of the positive charge on the sample, all XPS peaks in the spectrum shifts by the same amount to higher binding energies. Normally calibration is done using the binding energy of a known compound for example Si 2p in SiO_2 is taken as 103.4 eV or Carbon 1s binding energy is considered as 284.4 eV. There are other substances which can be added and also used for calibration.

In addition to the shifting of the peaks, the peaks can also broaden and thus decrease the resolution with a low signal to noise ratio. Charging in supported catalysts is usually limited to a few eV in normal mode of operation but when monochromatic XPS is used, it can be considerably more. These effects can be minimized by using a flood gun which showers low energy electrons on the sample or mounting the sample like powders in indium foils. It is also possible that the electrons generated by the source and heat produced can sometimes decompose the complex i.e., the organometallic complexes used as precursors for catalysts. In such cases one has to interpret the spectra obtained taking all these secondary effects that can occur inside the spectrometer like reduction of the sample in vacuum, the charging, the decomposition of the species due to heat and electrons and all other side effects due to operating conditions.

In supported catalysts, depending on the dispersion and the particle size of the supported phase, the intensity ratio of I_p/I_s due to the emission from the supported phase and the support, can be small when the support is exposed and the dispersion is small or can be large if the dispersion is good and the support phase is not exposed and covered by the supported phase.

This simple explanation will not hold well if the supported phase were to occupy and spread inside the pores of the support or the particles were to take different geometries, or if the support were to function as a stratified layer.

4.6 ANGLE DEPENDENT XPS

The intensity ratio of an over layer to the substrate is given by

$$I_0/I_s = \{(\sigma_0 n_0 \lambda_0(E_0))/(\sigma_s n_s \lambda_s(E_s))\}[1/\exp(-d/[\lambda_0(E_0)\cos\theta]/\exp(-d/[\lambda_0(E_s)\cos\theta]\}$$

where I_0 and I_s are the XPS intensities of over layer and substrate, n is the atomic densities in mol/cm^3, σ is the XPS cross section, usually taken from the tables of Scofield $\lambda(E)$ is the inelastic mean free path at the prevalent kinetic energy E through the over layer (λ_0) or the substrate (λ_s), θ is the take off angle with respect to the surface normal and d is the thickness of the over layers. Note that $\lambda_0(E_s)$ is the mean free path of electrons formed in the substrate traveling through the over layer. In the case that the over layers formed is a film of SiO_2 on a silicon single crystal the expression reduces to

$$I_{SiO_2}/I_{Si} = [\ n_{SiO_2}\ddot{e}_{SiO_2}\ 1\text{-}\exp(-d/\theta_{SiO_2}\cos\theta)]/[n_{Si}\ddot{e}_{si}\ \exp(-d/\lambda_{SiO_2}\cos\theta)]$$

The intensity ratio of Si^{4+}/Si is a function of take-off angle. The agreement up to take-off angles of about 60°C is good. At higher angles elastic scattering phenomenon reduces the intensity of the substrate electrons traveling at shallow angles from deeper regions.

4.7 ULTRAVIOLET PHOTOELECTRON SPECTROSCOPY

The commonly used sources are He I at 21.2 eV and He II light at 40.8 eV. Since at these low energies only valence electrons can be ejected, this spectroscopy is restricted to probing bonding in metals and molecules and in adsorbed states.

In Fig. 4.2 a schematic representation of a UP spectrum of a d-metal along with the density of states of its electrons. The onset of the UP spectrum is clearly seen at the Fermi level where the density of states is high and this provides a convenient zero for the binding energy. Most of the intensity at higher binding energies is due to secondary electrons. These are inelastically scattered photoelectrons from the valence band which have suffered energy loss while traveling through the solid. The shape of this secondary electron emission depends on the structure and composition of the material, the orientation of the sample with respect to the analyzer and the presence of other fields in the spectrometer.

It is clear that the electron coming from the Fermi level has a kinetic energy $E_k = h\nu - \varphi$ and the slowest electrons with $E_k = 0$ with the highest binding energy since the kinetic energy of the electrons in UPS is low between 5 to 15 eV and since the final state of the photoelectron is still in the region of the unoccupied part of the density of states of the metal this spectrum can be considered at joint Density of states of both occupied and unoccupied states. If the photon energy were to increase then the final states will be in the range corresponding to entirely free electrons in vacuum. If the photon energy increases then one can probe the true density of states of metals and in XPS also one probes the true density of states of metals but it is at a lower resolution due to broader line width of the X-ray source.

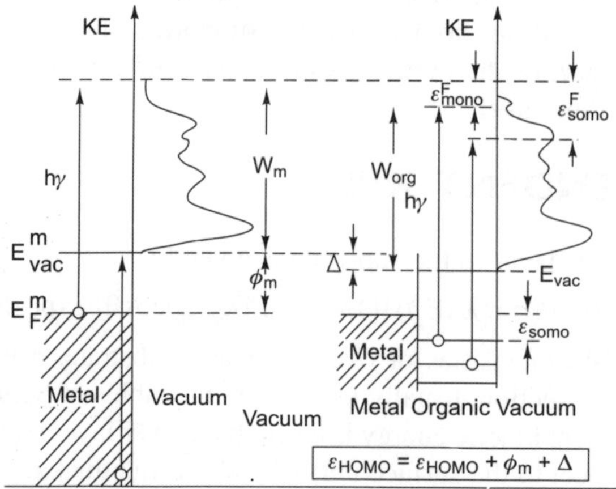

Fig. 4.2 Schematic representation of the photoelectron spectrum of Au (on the left) and with organic layer deposited on the metal substrate (in the right). It is seen that the density of state (DOS) of metal is probed by the photoelectron spectroscopy. (reproduced from Seki, Trans on Electron devices, 44 (1997) 1295)

UPS is most often used to study single crystals and to probe the molecular orbitals of chemisorbed molecules.

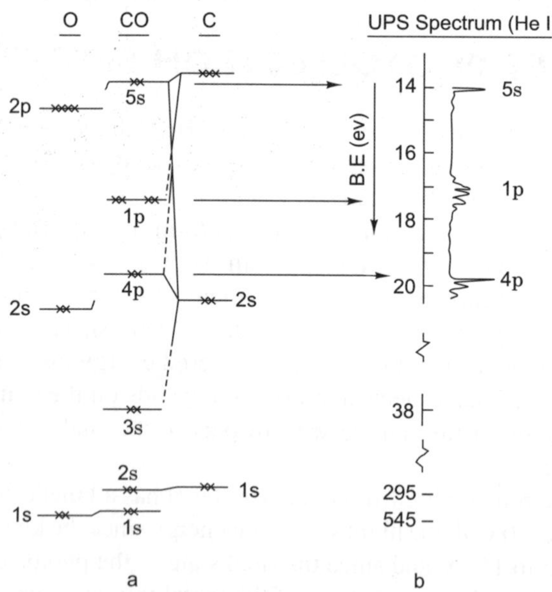

Fig. 4.3 The molecular energy diagram for CO and the corresponding ultraviolet photo electron spectrum (top half) and the UPS spectrum of CO adsorbed on nickel surface shown in the bottom half (*contd.*)

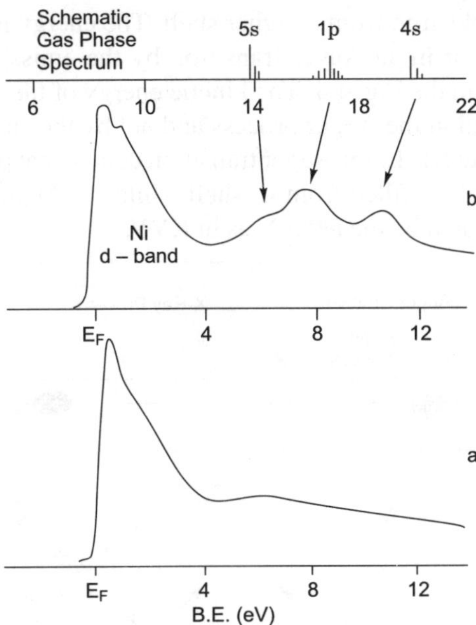

Fig. 4.3 The molecular energy diagram for CO and the corresponding ultraviolet photo electron spectrum (top half) and the UPS spectrum of CO adsorbed on nickel surface shown in the bottom half

The UP spectrum of silver layers deposited on ruthenium shows that the high density of d-state of Ru is seen till 5 eV below the Fermi level. However on deposition of silver these bands are attenuated while the d band of silver develops progressively between 1-4 monolayers until the energy distribution characteristic of silver is obtained for film thickness of about 10 monolayers. The effect of work function on the width of the spectra is also seen. Silver has a lower work function (4.72 eV) than ruthenium (5.52 eV) and the width of the spectra increases from 15.7 eV for Ru to 16.7 eV for a thick Ag film.

The adsorption of gases on single crystal metallic surfaces can be probed by UPS. One of the examples is the adsorption of CO on metallic surfaces. The MO level diagram and the corresponding UPS spectrum for gas phase CO is shown in the top of Fig. 4.3. On adsorption of CO on nickel surfaces, the molecular levels of are rearranged such that the 1π and 5σ levels overlap thus showing the orbitals involved in bonding with the metallic surfaces. Details of applications of Ultraviolet photoelectron spectroscopy will be dealt with in a separate chapter.

Xenon adsorption on Ru surface and Ag deposited ruthenium surface was studied. It was seen that the Xe 5p doublet at low coverages is at 6.7 eV for $5p_{1/2}$ which is characteristic of Xe on Ru (001). At higher coverages the Xe $5p_{1/2}$ peak occurs at 7.6 eV characteristic adsorption of Xe on Ag. The sequential adsorption on Ru and silver surfaces probably reflects that the adsorption energy of Xe on Ru is higher.

4.8 AUGER ELECTRON SPECTROSCOPY

Auger spectroscopy normally is carried out using electrons as the source in the energy of 1 KeV to 10 KeV . These electrons create core holes in the atoms of the sample. The excited atom relaxes by filling the core hole with an electron from a higher shell. The energy is either liberated as an X-ray photon (X-ray fluorescence) or in an Auger transition by the emission of a second electron, the Auger electron. As indicated in the Fig. 4.4. The kinetic energy of the Auger electron is determined by the electron levels involved in the Auger process and not by the energy of the primary electrons used. Three letter notations like KL_1L_2 for Auger transition denote that primary ionization takes place in the K shell and the core hole is filled from L_1 shell while the Auger electron is emitted from L_2 shell. Valence levels are indicated by the letter V as in KVV.

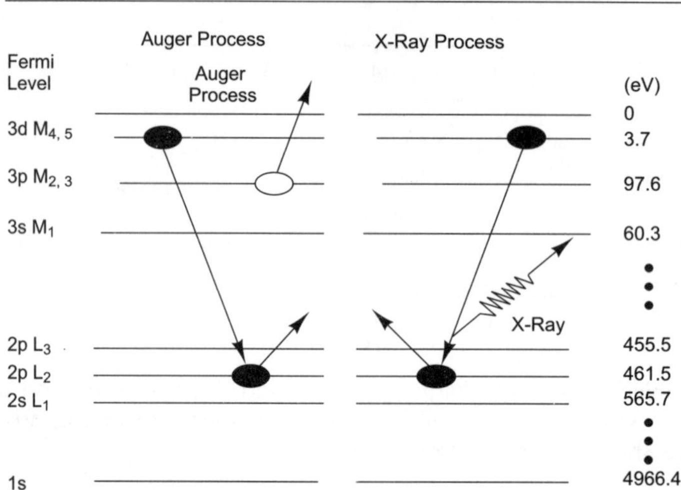

Fig. 4.4 Pictorial representation of Auger and X-ray processes

In the conventional electron excited spectrum, the Auger signal will be small against the background of secondary electrons originating from the primary beam. However, the spectra are usually shown in the derivative form.

The energy of Auger transition, for example for a KLM transition at a first approximation is given by

$$E_{KLM} \approx E_K - E_L - E_M - \delta E - \varphi$$

E_{KLM} is the kinetic energy of the Auger electron E_K is the binding energy of the electron in K-shell; E_L that of L shell and E_M that of M shell electron, is the work function and δE is the energy shift caused by relaxation effects due to relaxation taking place in the heavily excited doubly ionized atom.

Though chemical identity of the elements can be easily seen from Auger transitions, it is difficult to identify the oxidation states from Auger peaks as chemical shifts in XPS since relaxation processes are more complicated in Auger spectroscopy. It is only at times possible to get finger print data on the chemical nature of species.

4.9 INVERSE PHOTOEMISSION SPECTROSCOPY

Inverse photoemission spectroscopy (IPES) is the spectroscopy that measures the energy of photons ($h\nu$) emitted when electrons incident on a substance using an electron beam with a constant energy (E_i) relax to a lower energy unoccupied state (E_f). In this process, the energy conservation law is given by:

$$E_i = E_f + h\nu$$

By measuring E_i and $h\nu$, the unoccupied state (E_f) of the surface can be found.

Two modes are used for this measurement. One is the isochromat mode, which scans the incident electron energy and keeps the detected photon energy constant. The other is the tunable photon energy mode, which keeps the incident electron energy constant and measures the distribution of the detected photon energy. The latter can also measure the resonant inverse Photoemission spectrum. This is the technique which probes the unoccupied states, while all other electron spectroscopies probe the occupied states.

REFERENCE

1. A.W. Czanderna, Methods of Surface Analysis, Elsevier Scientific Publishing Company, 1975 pp.110-113.

Ultraviolet Photoelectron Spectroscopy (UV-PES)

In UV-PES, the sample is irradiated with light in the vacuum ultraviolet region of the spectrum. The photon source for these measurements is from a gas-discharge tube. Typical sources employed are given in Table 5.1.

Table 5.1 Typical sources employed in UV-PES

Source	Energy in eV
Helium I (doublet)	21.22
Helium II	40.81
Neon I (doublet)	16.7
Neon II	26.9

These energies are of the same order of magnitude as the binding energies for *valence shell electrons* of molecules and for the valence band states of condensed systems. This means that UV-PES gives information of only the valence orbitals in the molecular systems and possibly the valence band of solids. However, despite the energy limitation, UV-PES has advantages in many cases. The probability of photoemission varies with frequency. This can be characterized by the "absorption cross-section for photoemission". It turns out that this varies approximately as $v^{-3.5}$. Hence, excitation in the UV region is easier. In addition, the flux of photons produced by these discharge lamps is intense in comparison to the X-ray sources in X-PES. These factors mean that UV-PES can be more sensitive. It is therefore suited to working with dilute samples, such as gases (using differentially-pumped apparatus to obtain the low pressures required to make PES work, whilst still allowing there to be sufficient gas in the sample chamber) or to adsorbates on solid surfaces. UV-PES is also advantageous when working with delicate samples, as there is less chance of radiation damage. Finally, using UV-PES and varying the irradiating light allows one to see the frequency-dependence of the various PES peaks.

The investigations employing frequency dependence are particularly important for the study of solids. It can be shown on a quantum mechanical basis that the photoemission cross-section depends on the size, number of (angular) nodes and localization of the orbital in question. This means that by repeating the measurement of the PE spectrum with many different exciting radiation energies, one can identify the "d-orbital character" of the orbitals involved. This is useful for assigning bands in solid-state work. It is also an important test of the bonding model which is being used to explain the band structure of the solid in question.

UV-PE spectra can be employed to understand the molecular orbital picture of bonding in molecules. The two molecules of great relevance to catalysis are carbon monoxide and nitrogen. The molecular orbital scheme and the corresponding UV-PES spectra for nitrogen and carbon monoxide are given in Figs 5.1 and 5.2 respectively. It is seen that the HOMO level is a non-bonding (3σ) and the next orbital is strongly bonding (1π) and the third level is again a sigma orbital (2σ) in both the molecules. The UV-PE spectra do not show the deeply bound orbitals since the UP spectra were obtained with only He I light source which has a maximum energy of about 22 eV. It should be recognized that the lone pair electrons are localized on one atom namely C (3σ) and the other (2σ) on another atom, namely on oxygen. In the case of dinitrogen, the mixing of atomic orbitals results in 2σ and 3σ having little residual bonding character and they are the conventional "lone pairs" of dinitrogen.

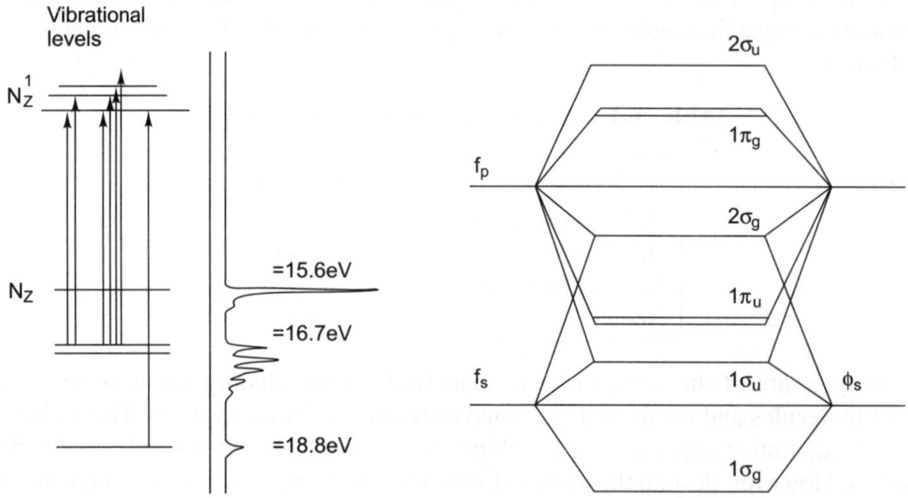

Fig. 5.1 (a) The UV photoelectron spectrum of N_2. The fine structure in the spectrum arises from the excitation of vibrations in the cation formed by photo-ejection, (b) the alternative ordering of orbital energies found in homonuclear diatomic molecules from Li_2 to N_2 (Note: Nitrogen atoms have first ionization energy of 14.5 eV)

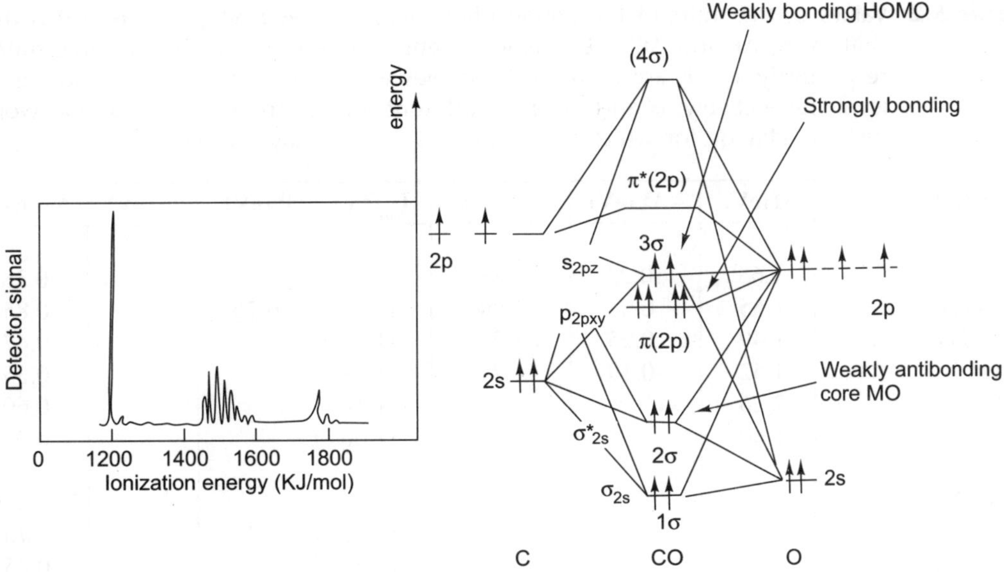

Fig. 5.2 The UV photoelectron spectrum of CO. The figure on the right is the molecular orbital diagram for CO

5.1 ADSORPTION OF GASES ON METALLIC SURFACES AS STUDIED BY UV-PES

It is known that the study of the electronic structure of adsorbates is important in understanding the bonding scheme in the adsorbed phase. PES is capable of providing information on the electronic structure of adsorption systems.

Table 5.2 Values of the shifts of the xenon photoelectron spectroscopic peaks (ΔI is the shift in eV, ΔE and ΔE^+ are interaction energies of the ground and ionic states respectively, I_{ads}, I^F are ionization energies with respect to vacuum and Fermi levels respectively, φ and $\Delta\varphi$ are work function for the clean metal and work function change for which coverage $\theta \to 1$ (monolayer coverage)

Substrate	ΔI (eV)	$-\Delta E$ (eV)	$-\Delta E^+$ (eV)	I_{ads} (eV)	I^F (eV)	Φ (eV)	$\Delta\varphi$ (eV)
(A_0)ML $\theta \to 0$							
Ni(100)	1.27	0.24	1.51	12.13	6.83	5.30	
Pd(111)	0.98	0.36	1.34	12.42	6.47	5.95	
Ag(111)	1.02	0.23	1.26	12.38	7.62	4.76	
W(110)	1.17	0.19	1.36	12.23	7.13	5.10	
Pt(111)	1.10	0.31	1.41	12.30	5.90	6.40	
Al(111)	1.18			12.22	7.74	4.48	

(Contd.)

Table 5.2 Values of the shifts of the xenon photoelectron spectroscopic peaks (ΔI is the shift in eV, ΔE and ΔE^+ are interaction energies of the ground and ionic states respectively, I_{ads}, I^F are ionization energies with respect to vacuum and Fermi levels respectively, φ and $\Delta\varphi$ are work function for the clean metal and work function change for which coverage $\theta \rightarrow 1$ (monolayer coverage)

Substrate	ΔI (eV)	$-\Delta E$ (eV)	$-\Delta E^+$ (eV)	I_{ads} (eV)	I^F (eV)	Φ (eV)	$\Delta\varphi$ (eV)
$(A_1)\ \theta \rightarrow 1$							
Ni(100)	1.65	0.23	1.88	11.75			0.38
Pd(111)	1.55	0.35	1.90	11.85	6.75		0.85
Ag(111)	1.49	0.25	1.74	11.91			0.47
W(110)	1.52	0.19	1.71	11.88			0.35
Pt(111)	2.26			11.14			0.60
Al(111)	1.80			11.60			0.29
(B) 2 ML							
Ni(110)	1.30			12.10	8.30		0.75
Pd(111)	0.83			12.57	7.47		0.85
Ag(111)	0.96	0.17	0.1–0.13	12.44	8.15		0.47
W(110)	0.99	0.14	1.13	12.41	7.66		0.35
Pt(111)	1.63	0.21	1.84	11.77			0.60
Al(111)	1.30			12.10	8.20		0.29
(C)							
CO/Ni(110)	0.80			12.50	6.30		−1.65
CO/W(110)	1.89			11.51	5.6		−0.55
C_2H_2/Pd(100)	1.27			12.13	7.28		0.80
O/W(110)	1.14			12.26	6.26		−0.90
C/Pt(111)	1.91			11.49			
(D_1) 3 ML							
Ni(110)	1.20			12.20	8.50		0.85
Pd(100)	1.04	0.19	1.23	12.36	7.80		1.09
Ag(111)	0.86	0.15	1.01	12.54	8.33		0.55
Al(111)	1.10			12.30			
(D_2) Multilayer							
Pd(100)	1.04			12.50			
(D_3)							
N_2/H_2O/Ni(110)	1.15			12.55	8.8		1.10
N_2/CO/Ni(110)	0.85			12.55	6.35		−1.65

Data reproduced from Ref. S. Ishi and B. Viswanathan, Thin Solid Films, 201 (1991) 373.

Since 1970, PES has been extensively used for the study of adsorption of various species on metallic surfaces. However the analysis of spectra of such systems is complicated because of the variations in the interaction between the adsorbate and the metals. It is normally expected that the analysis of the spectra of rare gas atoms adsorbed on metals is simpler because of the weak interaction between the adsorbate and the metal surfaces.

The UP spectra of xenon atoms adsorbed on metal surfaces show two peaks in the region 5-8 eV with respect to the Fermi level which are designated as μ_1 and μ_2. Typical UP spectra for the adsorption of xenon on W (110) at various coverages are shown in Fig. 5.3. The two peaks are usually assigned to the $^2P_{3/2}$ and $^2P_{1/2}$ levels of Xenon atom.

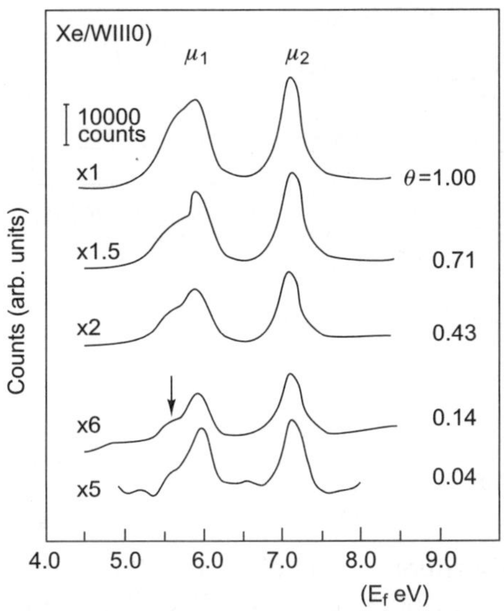

Fig. 5.3   UP spectra for the adsorption of xenon on W (110) at various coverages

The separation between these two peaks is 1.3 eV. In addition, the shape of the μ_1 peak is broadened more than that of the μ_2 peak with respect to those observed for gas phase xenon. However this simple explanation does not take into account the effect of the adsorption phenomenon itself. If one were to consider this bonding effect also, then the 5p orbitals of the xenon atom will be split into 5pσ and 5pπ orbitals for the case of interaction of the xenon with the metal surfaces. The resultant ordering of the split energy levels is the same as that obtained with the image potential model. Since the integrated experimental spectra are inadequate for deciding the ordering of the energy levels, the spin resolved photoelectron spectra (SRPES) is normally resorted to. One such measurement for Xenon adsorption on graphite is shown in Fig. 5.4. together with the similar excitation spectra as a function of exposure obtained with EELS (Electron energy Loss spectra) for the Xenon adsorption on gold.

The binding energy or ionization energy shift ΔI, which is defined as $\Delta I = I^v_{gas} - I^v_{ads}$ where I^v is the ionization energy with respect to the vacuum level, is rationalized in terms of an 'initial state' effect ΔI_i (an environment effect caused by the initial interaction) and of a 'final state' effect ΔI_f arising from the relaxation process taking place in the system after the ejection of the outgoing electron. The values of ΔI for a wide variety of substrates are summarized in Table 5.2 where the substrates are grouped in accordance with their different surfaces: (A) is the bare metal surface (A_0) when $\theta \to 0$, (A_1) when $\theta \to 1$; (B) is the second xenon layer;(C) is the first layer of species such as CO, C_2H_2, atomic oxygen and atomic carbon on metal surfaces; (D) is the third xenon layer (D_1) the xenon multilayer beyond the third layer (D_2) or the multilayer of molecular species (H_2O, N_2, CO) (D_3). It is seen from the values that the magnitude of ΔI is of the order of 1-2 eV irrespective of the nature of the metal surface.

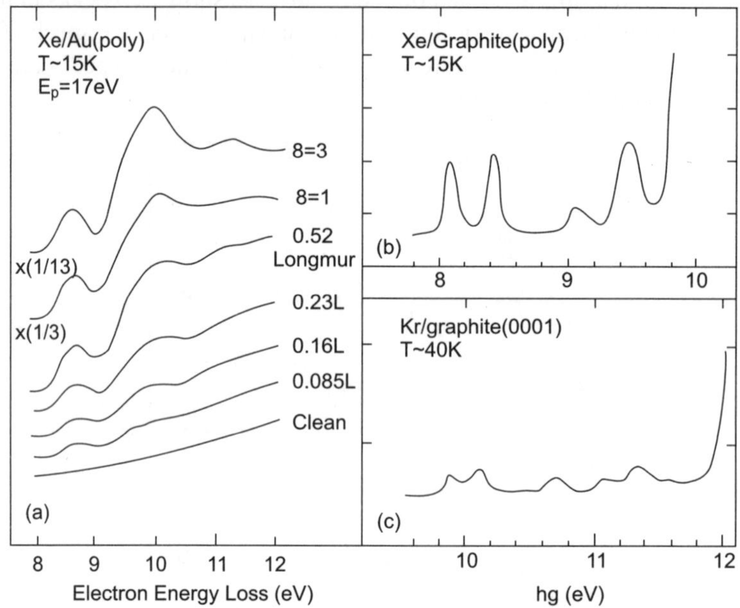

Fig. 5.4 SRPES Xenon adsorption on graphite together with the excitation spectra as a function of exposure obtained with EELS (Electron energy Loss spectra) for the Xenon adsorption on gold

When xenon atoms approach metal surfaces, it is expected that the degenerate 5p orbitals can be resolved owing to interaction with the metal surfaces. The electronic configuration of the ground state of a free xenon atom is ….$(5p)^6$. The 5p orbitals of xenon on adsorption give rise to two states designated as $5p\sigma$ and $5p\pi$. In this description one assumes that the metal possess a large number of electrons with suitable energy and symmetry for formation of molecular orbital and the metal levels in the ground state is filled up to the Fermi level. Therefore the electronic configuration of adsorbed xenon is given by ….$(5p\sigma)^2(5p\pi)^4$. the monoionic states possible for xenon have the configurations $(5p\sigma)^1(5p\pi)^4$ with a $^2\Sigma$ or $(5p\sigma)^2(5p\pi)^3$ $^2\pi$ State. The $^2\pi$ state can further split owing to spin orbit interactions into two degenerate states $^2\pi_{1/2}$ and $^2\pi_{3/2}$. The relative ordering of these two states can be either $E(^2\Sigma) < E(^2\Sigma)$ or the reverse depending on the magnitude of the interaction. This shows that the assignment of the peaks in UV-PES spectrum of adsorbed inert gases has to be treated where there is no lateral interaction between the adsorbed molecules.

5.2 PHYSICAL ADSORPTION OF DINTROGEN ON METAL SURFACES

The properties of dinitrogen in the physical adsorption state can be expected to be similar to those of the gas phase molecule. The UPS data for molecular nitrogen physisorbed on metal surfaces are given in Table 5.3.

The UP spectrum of physisorbed molecular nitrogen consists of three features due to $3\sigma_g$, $1\pi_u$ and $2\sigma_u$ valence orbitals. The peak separations and their relative intensities are the same as in the gas phase. All the energy levels are, however, shifted to lower binding energies by ~ 1-2 eV due to relaxation in the final state.

In the chemisorption of molecular nitrogen on metal surfaces, two states have been identified namely, the weakly adsorbed γ state (end-on orientation) and the strongly adsorbed α state (side-on orientation). The characteristic features of these states are summarized in Table 5.4. In the UV photoelectron spectrum the gamma state generally shows two features with the low energy feature containing both $3\sigma_g$ and $1\pi_g$ emission. The alpha state seems to exhibit a three peak spectrum as shown in Table 5.4.

Table 5.3 UPS data for the physical adsorption of molecular nitrogen on metal surfaces

Substrate	$3\sigma_g$ eV	$1\pi_u$ eV	$2\sigma_u$ eV	$(1\pi_u - 3\sigma_g)$ eV	$(2\sigma_u - 1\sigma_u)$ eV	T(K)
Gas phase dinitrogen	15.6	17.0	18.8	1.4	1.8	–
Cu (Condensed N_2)	11.1	12.5	14.3	1.4	1.8	~20
Ni flim	9.7	11.0	12.8	1.3	1.8	7
Ni(111)	9.2	10.5	12.3	1.3	1.8	45
Pd(111)	8.7	10.3	11.9	1.6	1.6	45
Al(111)	10.25	11.6	13.5	1.9	1.9	20

Data from Rao C.N.R and Ranga Rao G., Surface Science Reports, 13 (1991) 221.

Table 5.4 UP photoelectron spectra of molecular nitrogen chemisorbed on metals

Surface	$3\sigma_g$	$1\pi_u$	$2\sigma_u$	T(K)
γ state				
Ni(100)	7.6	–	12.4	77
Ni(111)	9.0	–	13.0	70
Fe(111)	8.6	8.3	11.8–12.2	<77
W(110)	7.0	–	11.7	100
α state				
Fe(111)	8.4	7.5	12.3	110
Cr(110)	8.4	7.1	12.7	90

Data extracted from Rao C.N.R and Ranga Rao G. Surface Science Reports, 13 (1991) 221

Composite bands due to $(3\sigma_g + 1\pi_g)$ are seen in most cases in the gamma state (weakly chemisorbed state of molecular nitrogen).

5.3 UPS STUDIES OF ADSORPTION OF CARBON MONOXIDE ON METALLIC SURFACES

Eastman and Cashion were the first to study the electronic structure of adsorbed CO on nickel surfaces by UPS measurements. Following this study a number of studies dealing with the electronic structure of adsorbed CO on metal surfaces are reported. Eastman and Cashion observed two peaks in the UP spectrum for CO adsorbed on nickel at ~7.5 eV and ~11.0 eV with respect to the Fermi level of the metal. They assigned these two peaks for emission from 3σ and 1π molecular levels respectively. This initial assignment was not correct. The correct assignment is that the low energy peak is a composite peak for emissions from 3σ and 1π while the peak at ~ 11.0 eV is due to emission from the 2σ level. It has been ascertained that in the adsorbed state the ordering of the energy levels is reversed and that the 1π has a lesser binding energy as compared to the 3σ level though in the gas phase CO the ordering is the reverse. The peak positions of the 3σ, 1π and 2σ levels of CO in the adsorbed state are at 7.84, 7.30 and 10.66 eV respectively. The observed peak positions are given for adsorption of CO on different planes of Group VIII metals in Table. 5.5.

Table 5.5 Peak positions of valence levels with respect to E_F and energy separation for $\tilde{2}\,\tilde{s}$ and $\tilde{1}\,\tilde{p}$ levels for CO adsorption on group VIII metals

Substrate	Peak positions				$\Delta(\tilde{2}\sigma - \tilde{1}\,\tilde{\pi})$
	$\tilde{3}\,\tilde{\sigma}$	composite	$\tilde{1}\,\tilde{\pi}$	$\tilde{2}\,\tilde{\sigma}$	
Fe(110)		7.5		10.4	2.9
Fe(100)		7.0		10.5	3.5
Fe(111)		7.6		10.8	3.2
Ru(0001)		7.6		10.7	3.1
Co(0001)		7.5		11.0	3.5
Rh(111)	8.3		7.8	11.2	3.4
Rh(100)		7.6		10.8	3.2
Rh(110)	7.6		8.7	10.6	1.9
Ir(111)	9.2		8.6	11.7	3.1
Ir(100)	9.2		8.6	11.5	2.6
Ni(111)	8.1		7.1	11.2	4.1
Ni(100)	8.3		7.8	10.8	3.0
Ni(110)	8.4		6.5	11.7	5.2
Pd(111)	8.2		7.5	11.2	3.7
Pt(111)	9.2		8.4	11.9	3.5
Pt(100)		8.8		11.4	2.6
Pt(110)	9.2		8.2	11.7	3.5
Co (gas)	14.0		16.9	19.7	2.8

Data adopted from reference, S.S. Ishi *et al.,* Surface Science, 161 (1985) 349

5.4 DETERMINATION OF THE VALENCE BAND STRUCTURE OF METALS

The UV-PES can also be employed to determine the electronic structure of metals and deduce the density of states of the valence band of metals. This is feasible since the emission starts from the Fermi level of the metals. In fact the UV PE spectra itself is referenced with respect to the Fermi level of the metallic systems. UV photoelectron spectra obtained for some typical metals like Ni, Cu, Zn and Au and CsAu are shown in Fig. 5.5.

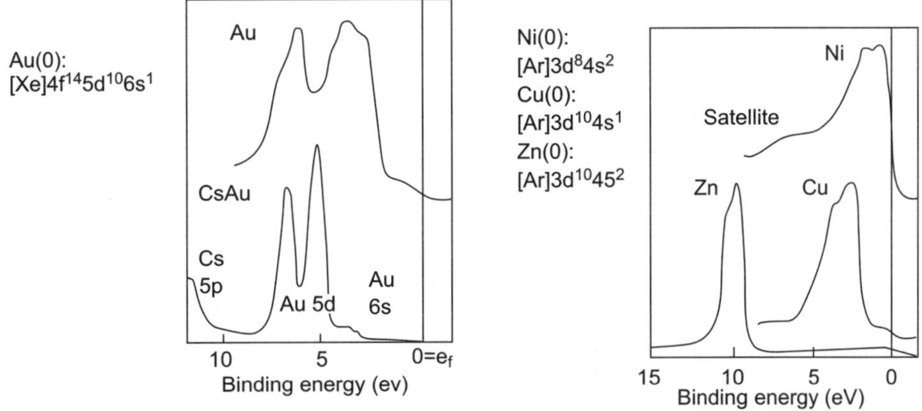

Fig. 5.5   UV-PE spectra of metals like Ni, Cu, Zn and Au and CsAu

The points that can be noted are:

1. The nickel valence band spectrum shows the satellite emission in the energy range around 5 eV which is a characteristic feature of this metal.
2. The copper valence band spectrum shows a shoulder on the higher energy side namely around 3 eV.
3. The valence band spectrum of zinc shows not only the higher energy shoulder but also the valence band shows higher binding energy sates.
4. The valence band spectrum of gold shows a split pattern which is also present in CsAu system.

In favourable cases the UV-PE spectrum of solids can be used to understand the electronic structure of valence band and also the density of states in this band.

REFERENCES

1. D.E. Eastman and J.K. Cashion, Phys. Rev. Lett., 27 (1971) 1520.
2. S.S. Ishi, Y. Ohno and B. Viswanathan, J. Sci. IInd.Res., 46 (1987) 541.
3. S.S. Ishi, Y. Ohno and B. Viswanathan, Surface Science, 161 (1985) 349.
4. S.S. Ishi and B. Viswanathan, Thin Solid Films, 201 (1991) 373.
5. C.N.R. Rao and G. Ranga Rao, Surface Science Reports, 13 (1991) 221.

Principle and Applications of Auger Electron Spectroscopy

The Auger effect was discovered independently by both Lise Meitner and Pierre Auger in the 1920's. Though the discovery was made by Meitner and initially reported in the journal Zeitschrift fur Physik in 1922, Auger is credited for the discovery in most of the scientific community.

The process can be observed with any of the probing beams like electrons, photons and ions. However, the practice is to use an electron beam with energies in the range 2-10 KeV. The process that one is considering is pictorially shown in Fig. 6.1.

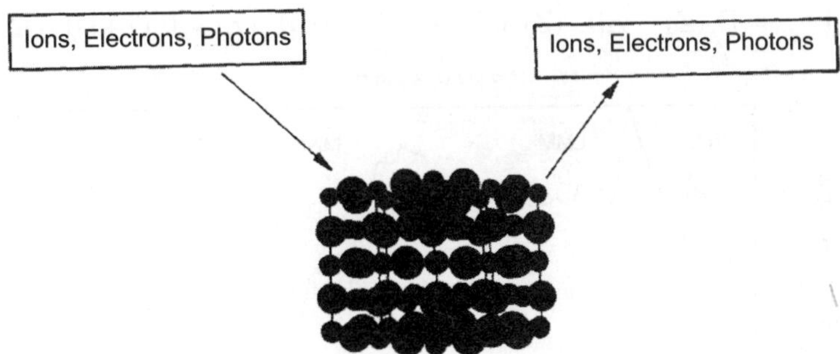

Fig. 6.1 Showing input particle beam and out-coming particle beams amenable for analysis

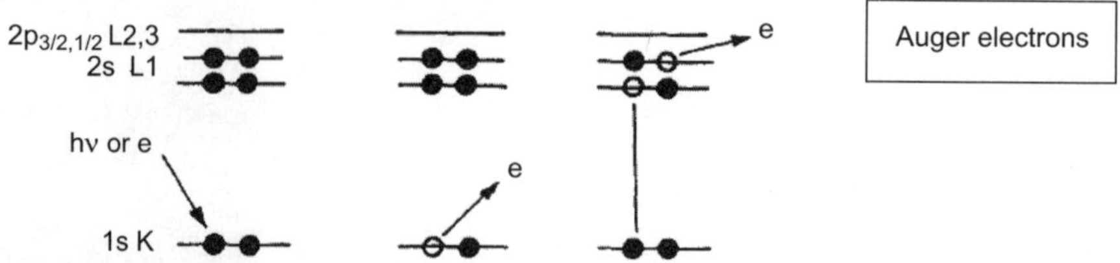

Fig. 6.2 Pictorial representation of the Auger process,[Primary core ionization, relaxation from an energy level above the core level and secondary ionization from an energy level energetically above the core that was ionized.]

In the Auger process a primary electron beam or a photon beam is employed to create a vacancy in the core energy level. The vacancy thus created in the core level is occupied by relaxation of an electron from another energy level. The energy thus released in the relaxation process can appear as a X-ray Photon (X-ray Fluorescence) or can be utilized to eject out a secondary electron (radiationless process) whose kinetic energy is measured. The process that is taking place is shown pictorially in Fig.6.2.

The kinetic energy of the auger electrons that is ejected out can be written to a first order approximation as

$$\text{Kinetic energy} = E_K - E_{L,1} - E_{L2,3} \tag{1}$$

Since three energy states are involved in the calculation of the kinetic energy, it is equivalent to probing three electronic states of elements and can hence be element specific. However, it should be remarked that there are ultimately two holes and two electron ejection, there must be some relaxation of the energies and one cannot simply substitute the energy values of the frozen states. In general the kinetic energy measured for an Auger electron can be written in simple approximation as follows:

$$E_{ABC} = E_A(Z) - E_B(Z) - E_C(Z+ \delta) - \Phi_s \tag{2}$$

Where, Φ_s is the work function of the material of the spectrometer. However even equation 2 is not the correct form for the kinetic energy of the Auger electron. Since in the Auger emission process one ends up with a doubly charged state as stated above we have to correct the atomic number (Z) suitably. This can be done in a variety of ways. One simple approximation can be written as

$$E_{ABC} = E_A(Z) - \tfrac{1}{2}[E_B(Z) + E_B(Z+1)] - \tfrac{1}{2} [E_c(Z) + E_c(Z+1)] - \Phi_s \tag{3}$$

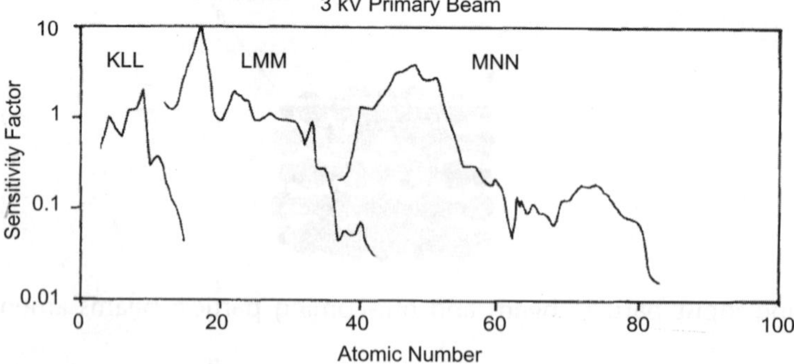

Fig. 6.3(a) Sensitivity factor as a function of atomic number for a primary electron energy of 3 keV.

It should be recognized that the expression is also one such approximation since the actual situation has to take into account the final state effects (coupling between the two unoccupied states) that will arise due to the emission of two electrons. The general guidelines for the observation of various Auger transitions are as follows: An Auger transition is generally characterized primarily by (1) the location of the initial hole generated and (2) the location of the final two unoccupied states that are generated. When describing the transition the initial core hole designation is lettered first, followed by the designations of the other two holes that are created in the decreasing order of order of the binding energy. For example the Auger transitions are designated as follows: KLL (for elements with Z = 3 to 14), LMM (for elements with Z = 14 to 40) and MNN (for elements with Z = 40 to 79 and MOO for other heavier elements. It should be remarked that in the solid state one can also observe transitions from the occupied valence band and these transitions are usually denoted as XVV (example KVV) where X stands for the core level ionization.

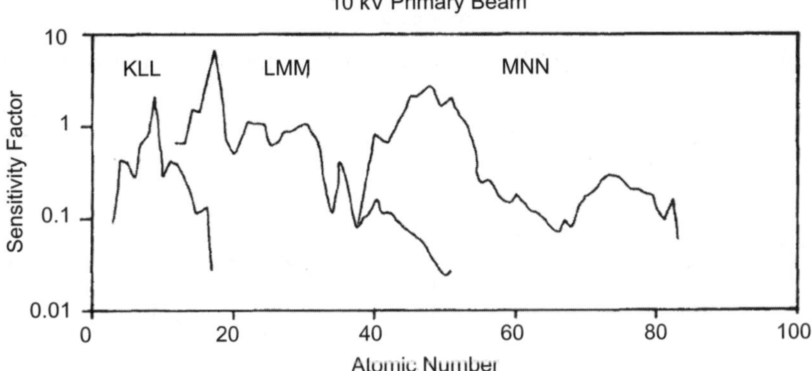

Fig. 6.3(b)   Sensitivity factor as a function of atomic number for a primary electron energy of 10 keV

It has been already stated that the energy of the primary electron beam can be varied between 2 to 10 keV. Accordingly the sensitivity factor for each of these transitions like KLL, LMM, MNN can also vary. In Fig. 6.3 the variations of the sensitivity factor for each of these transitions as a function of the atomic number are shown as a function of the atomic number for two primary electron energy namely 3 and 10 keV. It should be remarked that the kinetic energy of the Auger electrons are independent of the energy of the primary electron beam employed for the creation of ionization in the core level. In this sense, kinetic energies of Auger electrons differ from the kinetic energy of X-photoelectron since in XPS, the kinetic energy of the ejected photoelectron depends on the energy of the incident photons employed.

In a similar manner one can also use the kinetic energy of each of the Auger transitions to identify the element involved in the transition. The pictorial representation of this statement is given in Fig. 6.4.

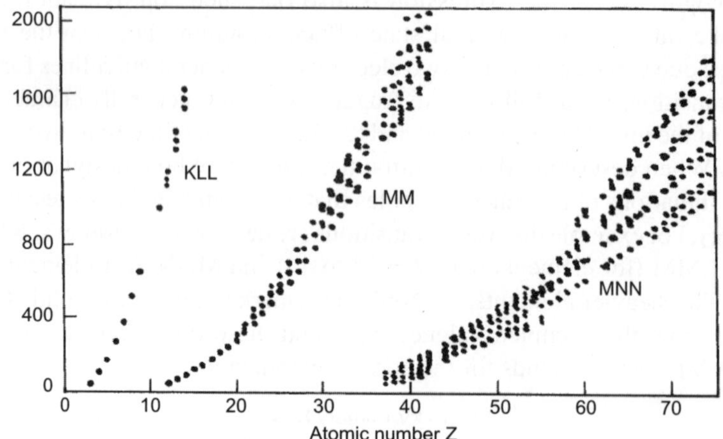

Fig. 6.4 Possible Auger transitions as function of atomic number of the element

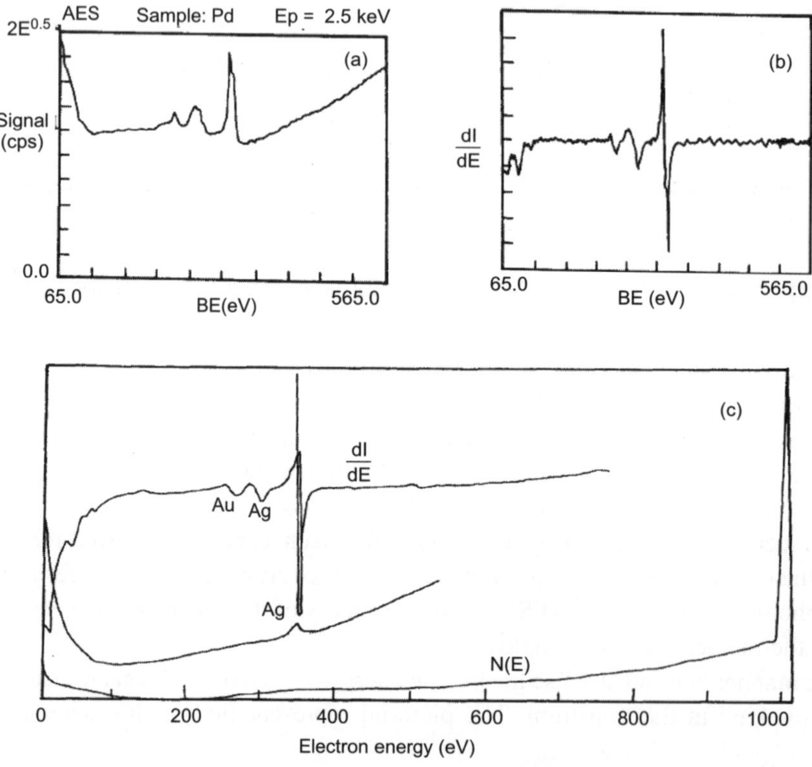

Fig. 6.5 (a) The Auger signal from Pd as function of the kinetic energy and it is seen that the signal appears as a small response in the raising background. (b) the signal in (a) is presented in the differential form and it is seen that the position of the signal in terms of the kinetic energy is clearly seen. (c) Auger signal for Ag is shown both in the same figure – the advantage of using differentiation is clearly seen

In Fig.6.5a the Auger signal generated using a 2.5 keV electron beam on Pd metal is shown. The main peaks for Pd occur between 220-340eV. The Auger signal is situated on a high background which arises from the vast number of so called secondary electrons generated by the multitude of inelastic scattering processes. It is therefore necessary to show the Auger spectrum in differential form. Since, in the differential form the signal is easily discerned and also is amenable for quantification. One such differentiation is shown in Fig. 6.5b for the same Pd Auger signal. Another example of the Auger signal of silver is shown both in the normal signal merged in the raising background as well as in the derivative form. It is clear that in the derivative form the signal is prominent. The position and intensity of the signals can be easily measured in the derivative spectrum.

6.1 DERIVING QUANTITATIVE DATA FROM AUGER ELECTRON SPECTRA

The possibility of obtaining quantitative data (at least relative values) makes Auger electron spectroscopic measurements more attractive compared to other surface analytical techniques. The Auger electron yield obtained is in the background of various secondary processes occurring simultaneously. Hence, obtaining absolute values for the concentration of chemical species may be difficult. However, relative concentration values can be extracted from the Auger spectra.

Table 6.1   Data extracted from Auger spectroscopic studies on first row transition metal nitrides to demonstrate how AES can be used as semi quantitative tool [reproduced from ref. *Surface Science Spectra*, **7**, 167-280, 2000]

Analysis	ScN	TiN	VN	CrN
Metal L3M2, 3M2, 3 Auger transition peak position (α) *(eV)*	337.0	384.2	435.4	486.8
Metal L3M2, 3M4, 5 Auger transition peak position (β) (eV)	367.2	417.4	472.0	527.8
Nitrogen K L2, 3L2, 3 Auger transition peak position (γ) (eV)	382.2(a)	(b)	382.4	381.6
Relative intensity for the as deposited samples Iα/Iγ	1.00	(b)	1.95	1.69
Relative intensity for the as deposited samples Iβ/Iγ	2.00	2.52(b)	1.43	1.30
Bulk composition (Rutherford Back scattering, RBS)	**1.06±0.03**	**1.02±0.02**	**1.04±0.02**	**1.02±0.02**
Relative intensity for ion bombarded samples Iα/Iγ	1.01	(b)	1.54	1.14
Relative intensity for ion bombarded samples Iβ/Iγ	1.82	2.10	1.01	0.94

(a) This nitrogen Auger peak overlaps with the metal Auger line L2M4, 5M4, 5 line; estimated that the metal contributes approximately 6%

(b) The nitrogen Auger peak has a severe overlap with Ti L3 M2, 3M2, 3 Auger line

The total Auger yield can be written by the following relationship

$$Y(t) = N_x \ X \ \delta t \ X \ \sigma \ (E, t)[(1- \omega_x] \exp(- t \cos \theta/\lambda) \ X \ I \ (t) \ X \ T \ X \ d(\Omega)/4\Pi \qquad (4)$$

where N_x is the number of X atoms per unit volume, λ is the electron escape depth, θ is the analyzer angle, T is the transmission of the analyzer, I(t) is the electron excitation flux at depth 't', $d\Omega$ is the solid angle and δt is the thickness of the layer that is being probed. It should also be noted that normally the Auger yield is greater for lighter elements as compared to the X-ray yield while the same is opposite for heavier elements and hence fluorescence yield is higher with heavier elements.

Since many of the terms in equation (4) are not known or difficult to determine accurately, most analyses compare measured yields with external standards of known composition. Ratios of the acquired data to standards can eliminate common terms, especially instrumentation characteristics and material parameters, and can be used to determine elemental composition. Comparison techniques work best for samples of homogeneous binary materials or uniform surface layers, while elemental identification is best obtained from comparison of pure samples.

One such quantitative analysis has been carried out for the first row transition metal nitrides namely, ScN, TiN, VN and CrN. The data that have been reported in literature is reproduced in Table 6.1.

The type of information that can be generated from Auger electron spectroscopic measurements that are relevant to catalysis include (i) Identification of elements; (ii) Identification and quantitative determination of elements on the surface; (iii) Depth profiling; (iv) Adsorption, desorption and surface segregation with respect to the bulk; (v) Chemical state of elements; (vi) Chemical reactivity measurements and (vii) Elemental mapping. In this chapter, examples some of these applications will be considered.

6.2 CHEMICAL STATE OF THE ELEMENTS

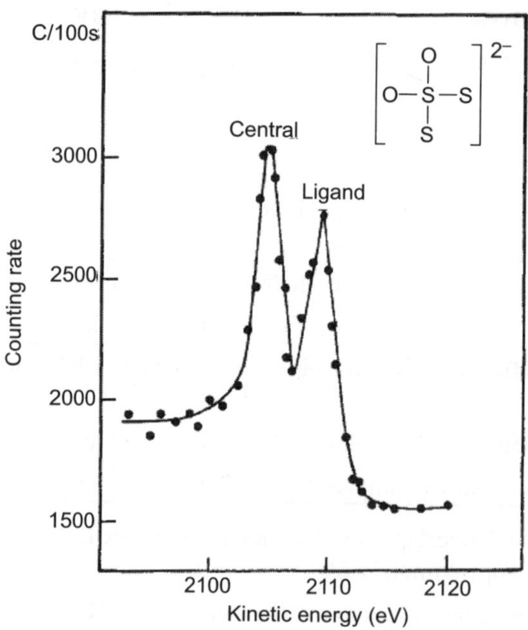

Fig. 6.6 The KLL transition from sulphur species from thiosulphate [Spectrum reproduced from the work of Fahlman et al., Phys. Lett., 20, 159 (1966)]

The valence state of the elements in the compounds can be deduced from the Auger electron spectroscopy. The KLL transition of sulphur in thiosulphate has been recorded and the spectrum obtained is shown on previous page in Fig. 6.6.

It is seen that two distinct signals for sulphur are seen the one at lower kinetic energy belongs to the central sulphur species (+6 Oxidation state) while the other is for the sulphur (–2 state) acting as the ligating centre. When core electrons are involved in the Auger process, it appears that the chemical shifts are similar to what is observed in XPS. The expected chemical shift in Auger line is given by $\Delta E = \Delta E_1 - 2\Delta E(L_{2,3}) \approx 5$ eV while the experimentally observed value is 4.7 eV.

It should be noted that the energy position of the peaks and at times the shape of the peaks also change with change in oxidation state or chemical environment of the species concerned. It is therefore necessary some times to compare with reference species for identification. As an illustration of this point the KLL and LMM transition of metallic aluminium and aluminium oxide are shown in Fig. 6.7. The possible chemical shift is shown in the Al LMM transition in this figure. However it must be remarked that the chemical shift is not as straight forward as it is in XPS due to final state effects arising out of two vacancies that are being created in Auger emission.

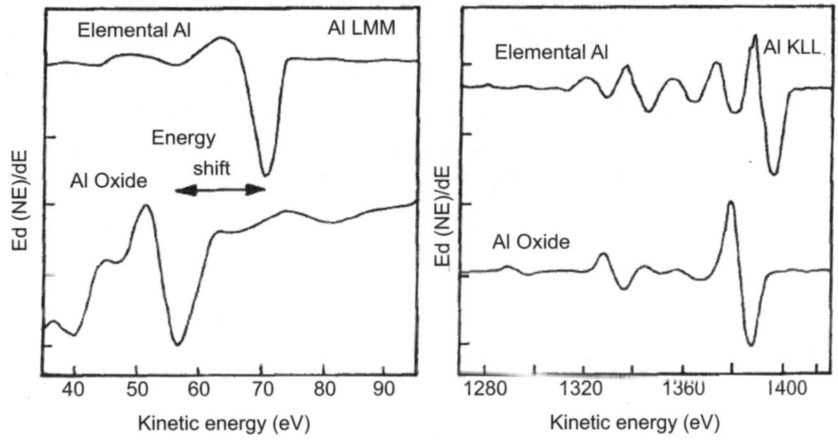

Fig. 6.7   The Auger emission of Al KLL and Al LMM for pure metal Al and Aluminum oxide to demonstrate the possible chemical shift as a result of oxidation state as well as how the analysis of Auger spectra are strengthened by way comparison with the pure element.

6.3 DEPTH PROFILING

Depth profiling is one of the prominent applications of Auger electron spectroscopy. This has been illustrated with a modeled system of multi-components shown in Fig. 6.8. On the left hand side a modeled configuration of the system of Al/Pd thin film on GaN substrate is shown. On the right hand side the depth profile of this system with ion bombardment is shown. One can easily see that the Al and Pd have been deposited on GaN substrate as it has been intended to be, that is Al on top of Pd layer and Pd layer in direct contact with GaN substrate. The oxygen signal on the top Al layer is due to exposure to air. If the same system was annealed one can visualize the diffusion of Al and Pd and hence the depth profile shows the presence of both these components.

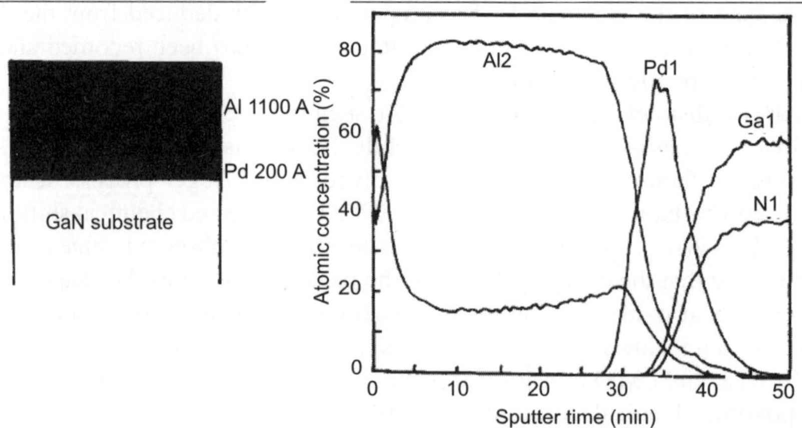

Fig. 6.8   Left Model of the thin film configuration of Pd and Al on GaN substrate (Right) Concentration profile for each of the element probed by Auger yield as function of sputter time with ions. The sputter time is proportional to the depth to which the system has been removed by ion bombardment

The high surface sensitivity of Auger electron spectroscopy is controlled by the mean free path of the electrons in the kinetic energy range normally encountered in AES namely 200 to 2000 eV. The Auger electron escape depth that have been evolved from various measurements to form the universal escape curve is shown in Fig. 6.9.

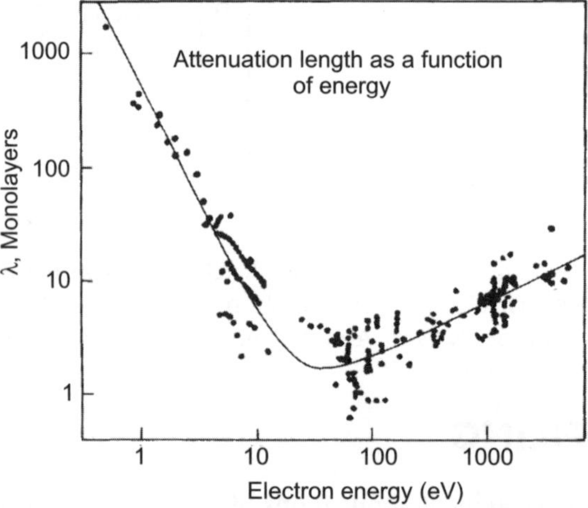

Fig. 6.9   Auger electron escape depth versus kinetic energy for electrons from various materials. The possibility of an universal curve indicates that the escape depth is independent of the material and matrix. This observation that the escape depth varies very little with respect to material makes the quantitative measurements with AES a possibility

6.4 ADSORPTION – DESORPTION STUDIES

Since high energy electron beam is used in Auger analysis, it is possible that the adsorbed state of the molecule may undergo changes and dissociation sometimes. Hence it is necessary one has to be careful to interpret the data obtained for adsorption of gases by AES. However, AES peak to peak height has been often used to calculate the coverage of the surface and the data generated are compared with the data obtained with Low Energy Electron Diffraction (LEED) or quartz microbalance measurements. A second parameter that is usually derived from Auger measurements is the value of the sticking coefficient as a function of coverage.

Oxygen adsorption on metallic surfaces at low temperatures is of great interest to understand the process of oxidation. In one such study on the adsorption of oxygen on Al (100) surfaces, the Auger signal intensities of the peak at 505 eV have been considered as a function of exposure of oxygen. These results show that for the initial amount of oxygen adsorbed non-stoichiometric oxide is formed and only after exposure to about 3.5 L (Langmuir exposure of 10^{-6} torr of the gas for one second), the familiar Al_2O_3 oxide is formed. A typical growth curve as a result of oxygen adsorption on Al(100) surface as monitored by AES is given in Fig. 6.10.

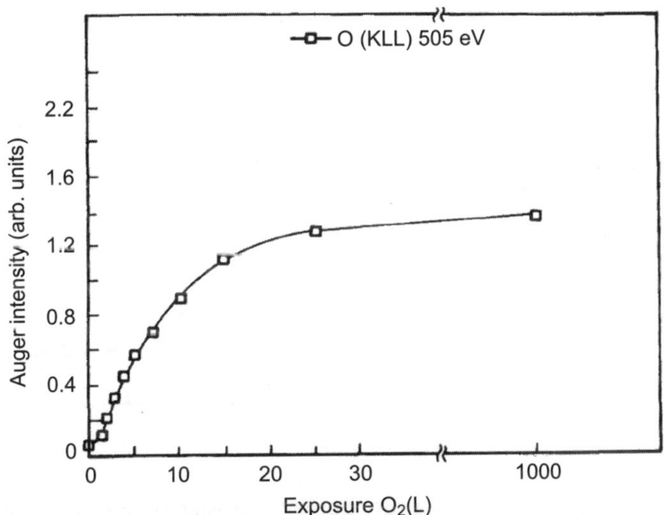

Fig. 6.10 Intensities of AES signals as a function of exposures to O_2 to Al (100) at 80 K. [A. Pashutski, A. Hoffman and M. Folman, Surf. Sci, 208 (1989) L91]

In Fig. 6.11, NOO Auger spectra of xenon adsorbed on Pd (100) are shown. The contributions from various layers are also shown. In Table 6.2 the available data for the shift in the 4d level as well as Auger transitions are given as a function of the extent of adsorption in terms of number of layers. It is seen that the shifts in core level binding energy and the shift in the Auger transition are higher for the first layer and for the subsequent layers the shifts are comparatively small.

Table 6.2 Binding energy shift for the 4d level and NOO Auger line shifts for the Xenon atoms adsorbed on Pd(100) with respect to the gaseous atoms

Layer	ΔI (4d) eV	Δ(Auger) of NOO line (eV)	Δ(Auger)/ΔI
First layer	2.14	6.57	3.07
Second layer	1.49	4.69	3.15
Outer layer	1.28	4.89	3.82

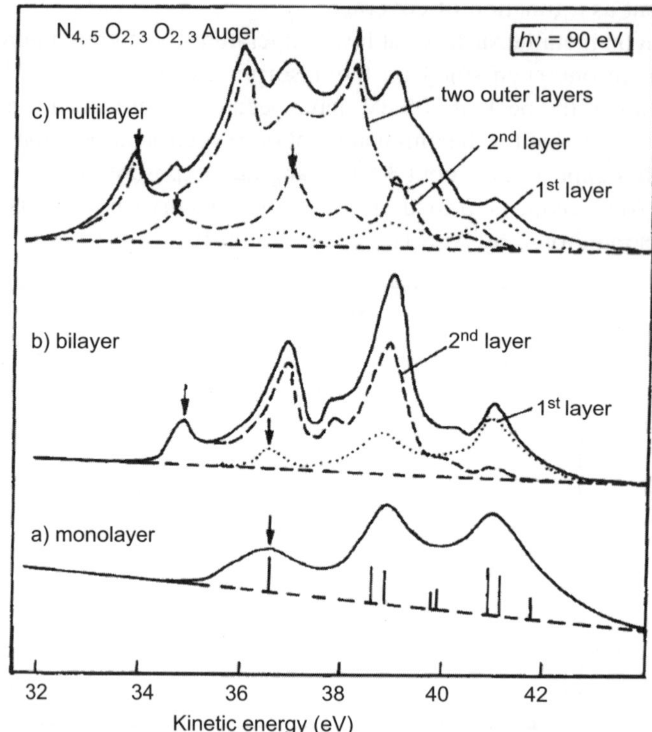

Fig. 6.11 NOO Auger spectra as a function of number of layers for Xe on Pd (100)

6.5 SURFACE REACTIONS

From an investigation of the rate of growth of the oxygen auger peak and the decay of the S peak as a function of temperature and partial pressures of oxygen, it has been deduced that the surface reaction consists of the following three steps.

$$O_2(gas) \;\rightarrow\; O_2 \;(ads)$$
$$O_2(ads) \;\rightarrow S \;(ads) \rightarrow SO_2 \;(ads) + nS \;(S = site)$$
$$SO_2(ads) \;\rightarrow\; SO_2(gas)$$

Kinetic studies of the growth of surface oxygen and the decrease of S have been used to conclude that the formation of SO_2 (ads) is the rate determining step and the reaction occurs by nucleation on a limited number of adsorbed S sites and then the reaction is spreading outward.

6.6 CATALYTIC ACTIVITY

Catalysis occurs on the surface or interface (in the case of electrochemical reactions) and hence it is required to probe the surface structure and the chemistry manifested by the surface species. Catalytic surfaces often absorb impurities and other species from the atmosphere and they act as poisons for the surface or occupy active sites thus preventing catalytic reaction. This situation is mostly true for noble metal catalysts. The identification of the presence of impurities and removal of these are important from the catalytic point of view. Essentially the extent of removal and the means of removal have to be ascertained. For example heating of Pt surfaces in oxygen atmosphere at 673 K can remove S and C impurities present on the surface.

In a similar manner, the oxidation of CO on Pt surfaces has been investigated and it has been shown that at low temperatures $373 < T < 473K$ CO remains adsorbed and the adsorption of oxygen is competitive to that of CO. An induction period was also observed for this reaction. However, at higher temperatures or on oxygen preadsorbed Pt surfaces, the oxidation of CO occurred instantaneously and the reaction was independent of the temperature. This type of mechanistic details could be studied by AES since it is capable of monitoring the surface species directly.

The formation of HCN in the oxidation of ammonia on Pt-Rh alloy catalyst has been identified to be due to the carbon impurities present even in sub-surface layers of the catalyst as identified by sputtering studies.

In the case of copper catalysts (used for dehydrogenation reactions) and Pt-alumina (used for hydrogenation of olefins) catalysts, it has been shown that lead and iron respectively act as poisons. The identification of the presence of these impurities on the catalyst surface has been carried out by AES.

Surface segregation and also reorganization are important in many catalytic systems especially that of alloy catalysts. For example in the case of Ni-Au alloy catalysts, the surface segregation of nickel has been observed in the presence of chemisorption of hydrogen or oxygen. This type of preferential accumulation at the surface can be due to various factors like, reaction induced migration, or due to the inherent thermodynamic control by the surface free energy of the species or due to some entropy factor.

Similarly, the reasons for the deactivation of the catalysts can be studied by AES. In the case of Ni-Al catalysts used for the methanation reaction, the preferential presence of S and preferential segregation of Ni and accumulation of S near nickel sites are the factors which contributed to the deactivation of the catalyst.

There are various other applications of AES especially in the evaluation of semiconductor materials and also to study metallurgical phenomena like corrosion, embrittlement in alloys, Grain boundary effects and wear. These applications have not been considered in this presentation.

6.7 COMPARISON OF THE TECHNIQUES

Since many of these techniques are being employed for the study of catalytic surfaces, it is necessary that a direct comparison is available. In Table 6.3 such a comparison is given.

Table 6.3 Comparison of the techniques for the applicability of surface analysis

Technique	Incident radiation	Emitted radiation	Property monitored	Elements detectable	Depth analysis	Spatial analysis	Identification Element/ Chemical species	Quantification possible/ not possible
AES	Electrons	Electrons	Energy	Li on	3-10μm	<12 nm	E & C	√
EDX	Electrons	X-rays	Energy	Be on	1 μm	1 μm	E	√
EELS	Electrons	Electrons	Energy	Li on	Depends	10 nm	E	0
ISS	Ions	Ions	Energy	Li on	Outer layer	100 μm	E	0
RBS	Ions	Ions	Energy	Li on	1 μm	1 μm	E	X
SIMS	Ions	Ions	Mass	All	1 nm	1 μm	C & E	X
XPS	X-rays	Electrons	Energy	He on	3-10μm	1 mm^2	E & C	√

[AES: Auger Electron Spectroscopy; EDX: Energy Dispersive X-ray; EELS: Electron Energy Loss Spectroscopy; ISS: Ion Scattering Spectroscopy; RBS: Rutherford Back Scattering; SIMS: Secondary Ion Mass Spectrometry; XPS: X-ray Photoelectron Spectroscopy]
{√ - good; 0 - reasonable; X - poor}

REFERENCES (Suggested Reading Material)

1. D. Briggs and M.P. Seah (Eds.) Practical Surface Analysis by Auger and X-ray Photoelectron Spectroscopy, New York, Wiley 1983.
2. K.D. Chids and others in C.L. Hedberg (Eds.) Handbook of Auger Electron Spectroscopy, Eden Prairie MN; Physical Electronics Publishing, 1995.
3. A.W. Czanderna in, Methods of Surface Analysis, Elsevier Scientific Company (1975).
4. A.W. Czanderna in S.P. Wolsky and A.W. Czanderna,(Eds.) Methods of Surface Analysis, Amsterdam, Elsevier, 1988.
5. G.E. McGuire and others in Sirface Characterization, Analytic chemistry, 65, 199R (1995).

Photo Electron Spectroscopy

7.1 INTRODUCTION

Photo emission from solids is one of the methods that can be advantageously used for the characterization of the solid substrates especially catalysts and their surfaces. The phenomenon of photo emission is one of the direct proofs of the Einstein's postulate. Three parallel endeavours led to the development of photo emission as a technique for characterization of solid state materials. They are: (i) Siegbahn and his group developed improved energy resolution of the electron spectrometers and combined it with X-rays [1-4]; (ii) Turner and his coworkers applied photoelectric effect to gases and by using ultra violet line from helium source; they could resolve vibrational fine structure of electronic levels [5-8]; (iii) Spicer measured photoelectron spectra from solid irradiated with UV light but the LiF window used to cut off all the photons with energies greater than 11.6 eV and hence he could only get a small part of the valence band [9-13]. Eastman and Cashion extended the energy range of about 40 eV by differential pumping and windowless UV sources [14-18]. The availability of synchrotron sources has extended the photon energy range further such that all energies between the ultra violet and X-ray regime are available. The mean free path of electrons decides the depth to which the material can be probed. The mean free path (λ) of electrons in elemental solids depends on the kinetic energy of electrons in the range 15–1000 eV and the mean free path for this range is ~ 0.5 nm and this is an ideal situation for surface sensitivity.

The zest to understand the surface chemistry such as surface chemical bonding and catalysis, made one to probe the types of atoms present on the surfaces, the phenomenon of adsorption, adsorption sites on the surfaces, adsorbed species and their bond strengths and the surface reactivity. All these properties can be probed by surface spectroscopic techniques. For the past 30 years, this type of spectroscopy has emerged as a key tool in surface analysis, mainly for:

1. Quantitative analysis
2. Information on the chemical nature and state of the detected elements

Electron spectroscopic studies have profound importance in the field of catalysis as it gives a thorough investigation of the electronic states of the elements concerned in any compound in any form under vacuum.

7.1.1 Advantages and Limitations of Electrons for Surface Spectroscopy

The main advantages of surface spectroscopy are:
1. The escape depth of electrons is only a few angstroms. Electrons are therefore suited to probe electronic states in the surface region of a sample (e.g., surface states).
2. Electrons are easily focused and their energy is easily tunable by electric fields.
3. Electrons are easy to detect and to count.
4. Using electrostatic fields it is easy to analyze the energy and the angle of an electron (thus determining the momentum). These instruments can be built small and handy, so that it can easily be moved (e.g., in angle resolved studies).
5. Electrons vanish after they have been detected. This is an appreciable advantage compared to ions or atoms, which are also used in surface physics.

Working with electrons involves some disadvantages also. They are:
1. Ultra high vacuum is necessary for these experiments.
2. The small escape depth of the electrons sometimes makes it difficult to distinguish between surface and bulk properties of the sample.

A photon impinging on the sample liberates an electron via the photoelectric effect and this electron escapes into vacuum. The energy of the incoming photon can be in the ultraviolet regime (5 to 40 eV, UPS), in the soft X-ray regime (100 to 1000 eV, SXPS) or in the hard X-ray regime (> 1000 eV, XPS). Photoelectron spectroscopy utilizes photo-ionization and energy-dispersive analysis of the emitted photoelectrons to study the composition and electronic state of the surface region of a sample. In general, XPS is used for the analysis of the surfaces of the solids whereas UPS generally pertains to the gaseous samples.

7.1.2 UHV Requirement for Surface Spectroscopic Studies

When one considers solid surfaces with atomic description (not polycrystalline and pressed pellets), they need to be cut carefully to expose a particular crystal plane and maintain atomically clean during the experiment (20-100 minutes). The crystal surface gets contaminated due to the bombardment of reactive gas molecules. Therefore the crystals have to be kept under ultra high vacuum during experimental surface analysis. Due to the availability of vacuum technology, it is now possible to create and maintain routinely various degrees of vacuum. It can be classified as

Rough (low vacuum) : $1\text{-}10^{-3}$ Torr (remove adsorbed gases from the sample)
Medium vacuum   : $10^{-3} - 10^{-5}$ Torr (eliminate adsorption of contaminants on the sample)
High vacuum   : $10^{-6} - 10^{-8}$ Torr (prevent arching and high voltage breakdown)
Ultra high Vacuum   : $< 10^{-9}$ Torr (increase the mean free path for the electrons, ions and photons)

UHV is also required to conduct low energy electron and ion based experiments without undue interference from gas phase scattering. From Kinetic theory of gases, the rate of surface bombardment by molecules or collision number (Z) is given by

$$Z = P / (2 \times \pi \times m \times k \times T)^{1/2}$$

where P = ambient pressure in Pascal, m = molecular mass in Kg, T = absolute temperature (K), and k = Boltzman constant. Assuming the sticking probability as 1, one can estimate the value of Z for typical gaseous contaminant CO, at 300K and 10^{-6} Torr as 3.82×10^{14} molecules per seconds per cm^2.

Typical atomic density of surfaces is of the order of 10^{15} cm^{-2}. Therefore the contamination rate can be calculated as 0.381 monolayers s^{-1} and the time required to form one monolayer at 10^{-6} Torr is 2.6 s. At P $= 10^{-10}$ Torr, the formation of one monolayer takes several hours and reliable surface science experiments can be performed within this time.

Normally the unit for UHV exposure is the Langmuir (L), where 1 L = 10^{-6} Torr s^{-1}

Sample surfaces in vacuum chamber can be cleaned by argon ion bombardment which is essentially the physical removal of surface material in layers. The energy of the argon ions used is in the range 100-3000 eV. After bombarding the surface with high energy ions, the surface needs to be annealed at temperatures close to the melting point to remove the surface defects [19].

Surface science experiments are carried out in high vacuum environment not only to keep the surfaces clean, but also to have the surface spectroscopic techniques operated as well. Surface spectroscopic techniques involve ions, atoms, or electrons whose mean free path in the vacuum environment must be greater than the dimensions of the apparatus so that these particles may travel to the surface and then to the detector without undergoing any interaction with residual gas phase molecules. In order to have reliable clean surfaces, the background gas pressure must be kept at minimum. For more reactive surfaces, it should be cleaned by physically removing the material from the surface in the preparation chamber.

7.1.3 Basic Principle and Instrumentation

The basic concept of photoelectric effect is that irradiated surfaces eject electrons which will carry information of the surface from where it is being ejected. This forms the basis of the photoelectron spectroscopy.

The energy of a photon is given by the Einstein's relation:

$$E = h\nu$$

where h is Planck constant (6.62×10^{-34} J s) and ν is the frequency (Hz) of radiation. Photoelectron spectroscopy uses monochromatic sources of radiation. In XPS, the photon is absorbed by an atom in a molecule leading to ionization and the emission of a core (inner shell) electron. By contrast, in UPS the photon interacts with valence levels of the molecule leading to ionization by removal of one of these valence electrons. The kinetic energy distribution of the emitted photoelectrons (i.e., the number of emitted photoelectrons as a function of their kinetic energy) can be measured using any appropriate electron energy analyzer and a photoelectron spectrum can thus be recorded.

One way to look at the overall process is:

$$A + h\nu = A^+ + e^-.$$

Conservation of energy then requires that:

$$E(A) + h\nu = E(A^+) + E(e^-).$$

Since the energy of electron is comprised solely as kinetic energy (KE) this can be written by the following expression.

$$KE = h\nu - (E(A^+) - E(A)).$$

The final term in brackets, represents the difference in energy between the ionized and neutral atoms and is generally called the binding energy (BE) of the electron. This then leads to the following commonly quoted equation:

$$KE = h\nu - BE$$

The electron being ejected has to surmount the work function of the sample under study, to come out with certain amount of kinetic energy. If ϕ is the work function, the above equation can be written as $KE = h\nu - BE - \phi$. For each and every element, there will be a characteristic binding energy associated with each core atomic orbital i.e., each element will give rise to a characteristic set of peaks in the photoelectron spectrum at kinetic energies determined from the photon energy and the respective binding energies. The presence of peaks at particular energies therefore indicates the presence of a specific element in the sample under study. Furthermore, the intensity of the peaks is related to the concentration of the element within the sampled region. Thus, the technique provides a quantitative analysis of the surface composition and is sometimes known by the alternative acronym, ESCA (Electron Spectroscopy for Chemical Analysis). Normally the sample and the spectrometer are in electrical contact if the sample is conducting. The Fermi level of these two systems will be the same. If an electron were to be ejected from the substance, say, a metal from the Fermi level, it is still bound to the metal. However the electrons coming from the Fermi level of the sample are detected with kinetic energy equal to $h\nu - \phi_{sp}$ where ϕ_{sp} is the work function of the spectrometer and not that of the sample. Removal of an electron from the Fermi level of the sample will require ϕ, the work function of the sample. The liberated electron is now at the potential outside the sample but not at the absolute vacuum level yet. Thus on the way to the analyzer, the electron is accelerated or decelerated by the work function difference which is the contact potential between the sample and the spectrometer [20].

The Basic Requirements for a Photoemission Experiment (XPS or UPS) are:

1. A source of fixed-energy radiation (an X-ray source for XPS or, typically, a He discharge lamp for UPS).
2. An electron energy analyzer (which can disperse the emitted electrons according to their kinetic energy, and thereby measure the flux of emitted electrons of a particular energy).
3. A high vacuum environment (to enable the emitted photoelectrons to be analyzed without interference from gas phase collisions) [20].

7.1.3.1 The Basic Instrument

The schematic diagram of a typical spectrometer is shown in Fig. 7.1. It essentially consists of X-ray, UV and electron beam sources, sample preparation and analysis chambers, analyzer and detector systems. All these systems are maintained under ultra high vacuum (UHV).

The development of synchrotron radiation sources has enabled high resolution studies to be carried out with radiation spanning a wider and more complete energy range (5-5000 eV). It has a number of desirable properties as listed below:

1. A continuous spectral distribution from the IR region to the X-ray region
2. High intensity
3. A high degree of collimation
4. A high degree of polarization

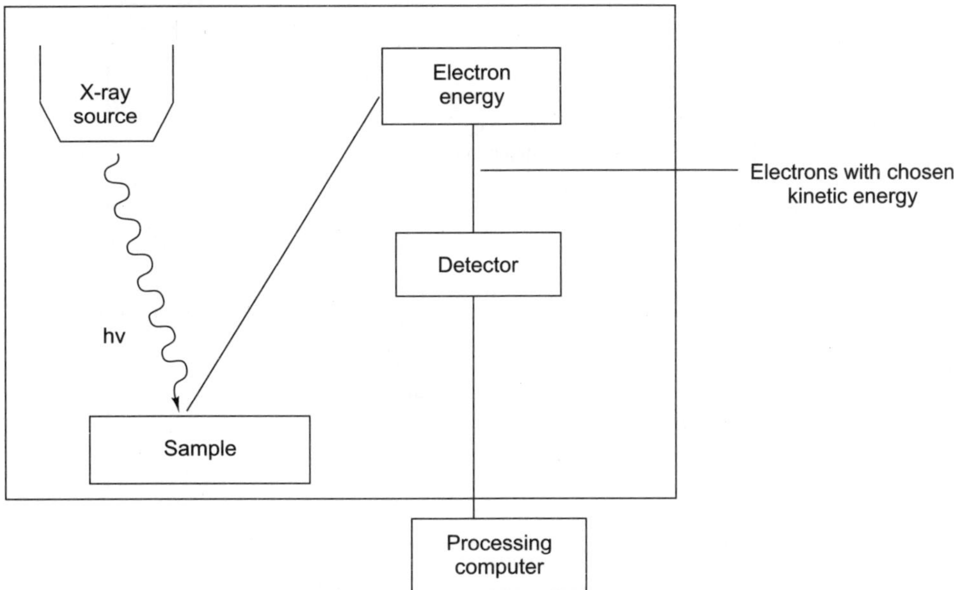

Fig. 7.1 Schematic arrangement of the basic elements of an X-ray photoelectrometer

An important source of both far ultraviolet and X-ray radiation is the storage ring of synchrotron radiation source (SRS). An electron beam is generated in a small linear accelerator. The pulse of these electrons are injected tangentially into a booster synchrotron where they are accelerated by 500 MHz radiation which also causes them to form bunches with a time interval of 2.0 ns between bunches. The electrons are restricted to a circular orbit by dipole bending magnets. When the electrons have achieved an energy of 600 meV they are taken off tangentially into the storage ring where they are further accelerated to 2 GeV. Subsequently only a small amount of power is needed to keep the electrons circulating. The stored orbiting electrons lose energy continuously in the form of electromagnetic radiation which is predominantly plane polarized and emerges tangentially in pulses of length 0.17 ns and spacing 2 ns. The radiation is continuous from the far infra red to the X-ray region but the gain in intensity, compared to conventional sources, is most marked in the far ultra violet and particularly, the X-ray region where the gain is of a factor of 10^5–10^6. A monochromator must be used to select a narrow band of wavelengths.

But such a work is, and will remain, a very small minority of all photoelectron studies due to expense, complexity and limited availability of such sources.

The most commonly employed X-ray sources are:

Mg K_α radiation: hv = 1253.6 eV

Al K_α radiation: hv = 1486.6 eV

Comparison of x-ray photoelectron spectra from copper using both Al K_α and Mg K_α is given in the Fig. 7.2.

The emitted photoelectrons will therefore have kinetic energies in the range of 0-1250 eV or 0-1480 eV. Since such electrons have very short lifetimes in solids, the technique is necessarily surface sensitive.

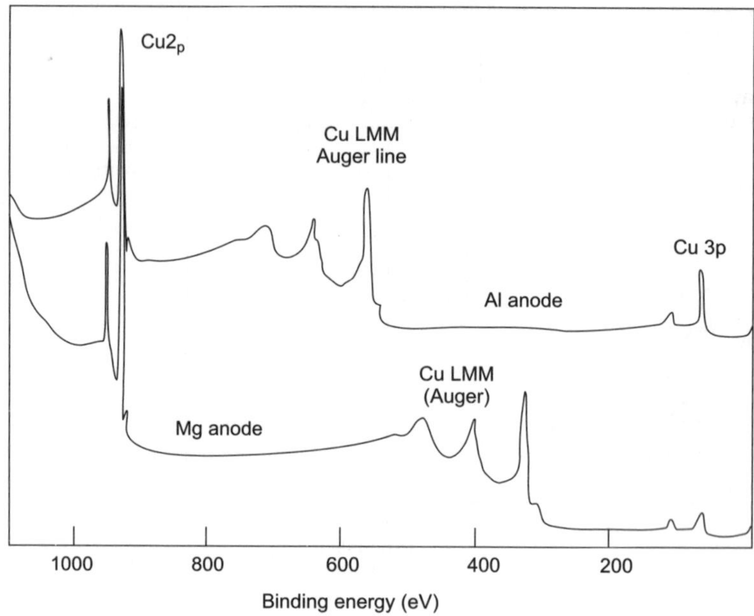

Fig. 7.2 Comparison of XPS spectra from copper using Al Ka (upper) and Mg Ka (lower) radiation. (reproduced from ref. 22)

One can realise that on the BE scale of the XPS peaks remain at constant values for both the sources, but the X-AES transitions are shifted by ± 233 eV on switching between the two sources. The commonly used sources for Ultraviolet photoelectron spectroscopy are He I at 21.2 eV and He II line at 40.8 eV. For Auger electron spectroscopy electrons or X-rays of energy 1-10 keV are used as the common sources.

7.1.4 XPS Spectra and Description

In XPS spectrum, one measures the intensity of photoelectrons N(E) as a function of their kinetic energy or more often in terms of binding energy. The abscissa for photoelectron spectroscopy is the kinetic energy with its zero at the vacuum level of the sample (KE = $h\nu$ – BE – ϕ)

Experimentalists generally prefer to use BE as the abscissa. BE is the so-called binding energy of the electrons which in solids is generally referred to the Fermi level and in free atoms or molecules to the vacuum level. Fig. 7.3 shows a XPS spectrum obtained from a Pd metal sample using Mg K$_\alpha$ radiation.

The photoelectron peak designations are based on the quantum numbers of the levels from which the photoelectron is originating. Spin orbit splitting as well as binding energies of a particular electron energy level increases with increase in atomic number.

1. The valence band (4d, 5s) emission occurs at a binding energy of 0-8 eV (measured with respect to the Fermi level or alternatively at 4-12 eV if measured with respect to the vacuum level).
2. The emission from the $4p$ and $4s$ levels gives rise to very weak peaks at 54 and 88 eV respectively.
3. The most intense peak at 335 eV is due to emission from the $3d$ levels of the Pd atoms, whilst peaks at 534 and 561 are due to emission from $3p$ and the peak at 673 eV is due to the emission from $3s$ level.

4. The remaining peak is not an XPS peak at all! - It is an Auger peak arising from X-ray induced Auger emission. It occurs at a kinetic energy of 330 eV (in this case it is really meaningless to refer to an associated binding energy).

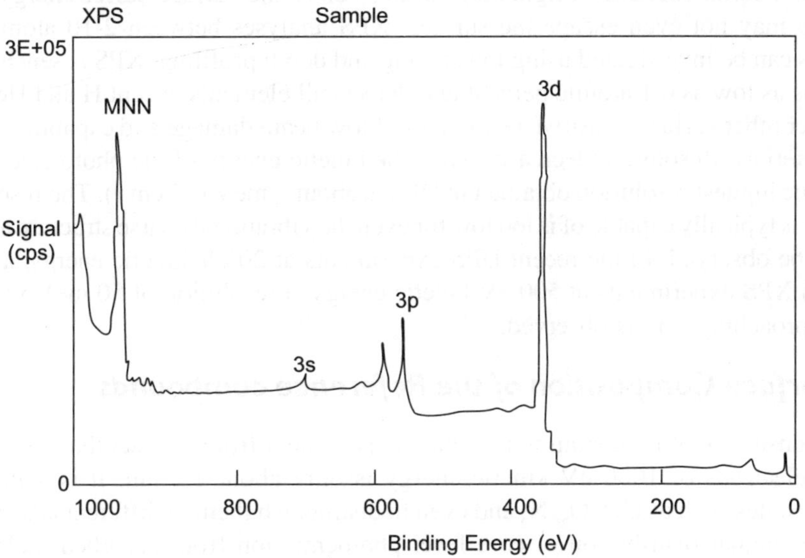

Fig. 7.3(a) X-ray photoelectron spectrum of Pd sample using Mg K$_\alpha$ line (ref. 20)

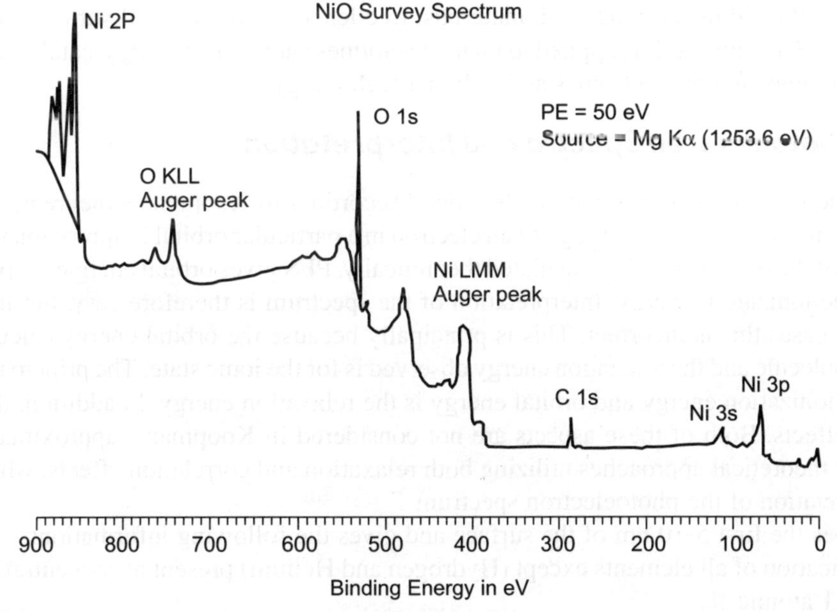

Fig. 7.3(b) Typical survey spectrum of NiO (one can see the emission from prominent Ni levels and also the emissions from Auger transitions)

7.1.4.1 Resolution of the Spectrometer

XPS is a surface-sensitive technique as the mean free path of photoelectrons is of the order of a few nanometers in a solid. Electrons originating further below the surface suffer energy loss through collisions and may not even escape the surface. XPS analyses between 2-10 atomic layers, but greater depths can be investigated using ion etching and depth profiling. XPS is sensitive to surface concentrations as low as 0.1 atomic percent and detects all elements except H and He. It offers the advantage over other surface-sensitive techniques of low beam damage and capability of analysis of insulating materials. Resolution decreases when the kinetic energy of the photo electrons is below about 5 eV. The highest resolution obtained in UPS is about 4 meV (32 cm^{-1}). The resolution 0.2 eV of which XPS is typically capable of is too low for even the vibrational coarse structure accompanying ionization to be observed. In the recent UPS experiments at 20 eV kinetic energy, a resolution of 3 meV and in XPS experiments at 500 eV kinetic energy, a resolution of 50 meV with an angular resolution approaching 0.1° is observed.

7.1.4.2 Surface Composition of the Reference compounds

The surface sensitivity of photoemission technique is evident from the fact that the inelastic mean free path of electrons of 1000 eV kinetic energy is only about 1.2 nm. It is suitable to detect adsorbed molecules such as CO, O_2, N_2 and even to distinguish them in different adsorption sites on surfaces. The angular distribution of core level photoemission from adsorbed molecules can be analyzed to determine the orientation of molecules on the surfaces. Currently photoelectron spectroscopy has been employed to characterize surfaces of variety of materials that include metals, oxides, alloys, thin films, polymers, biomaterials, tree leaves, ceramics and various single crystal catalysts by *in situ* methods. It is applied to many disciplines such as metallurgy, catalysis, corrosion, adhesion, tribology, micro electronics and polymer technology.

7.1.4.3 Photoelectron Spectra and Interpretation

Ionization energies of valence and core electrons: According to Koopman's theorem, for a closed shell of molecule, the ionization energy of an electron in a particular orbital is approximately equal to the negative of the orbital energies calculated theoretically. PES gives orbital energies experimentally and hence the ionization energy. Interpretation of the spectrum is therefore easy, but it turned out that in many cases this is incorrect. This is principally because the orbital energy calculated is for the neutral molecule and the ionization energy observed is for the ionic state. The principal difference between the ionization energy and orbital energy is the relaxation energy. In addition, there can be correlation effects. Both of these aspects are not considered in Koopman's approximation. There can be better theoretical approaches utilizing both relaxation and correlation effects, which leads to a full interpretation of the photoelectron spectrum.

XPS probes the first 5-10 nm of the surface and gives the following information:

1. Identification of all elements except (Hydrogen and Helium) present at concentrations greater than 0.1 atomic %.
2. Semi-quantitative information (± 5 %).
3. Molecular environment (Oxidation state, bonding with other atoms).
4. Lateral variations in surface compositions.

5. Non-destructive elemental depth profiles and surface heterogeneity (10 nm).
6. Destructive elemental depth profiles (1 µm).

7.2 FACTORS COMPLICATING THE XPS SPECTRA

7.2.1 Spin-orbit Splitting

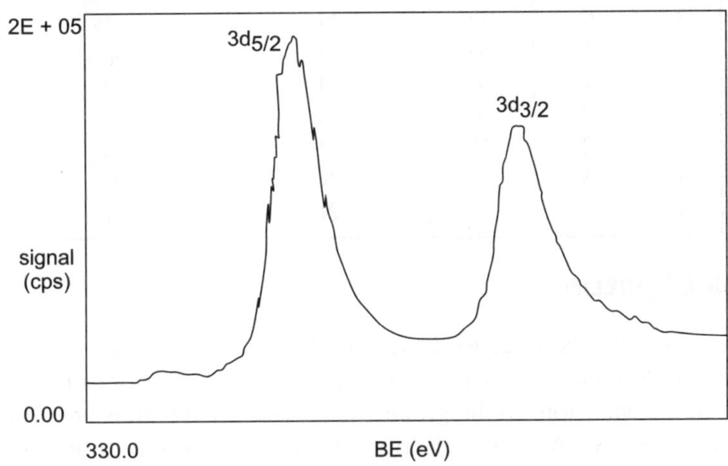

Fig. 7.4 Spin-orbit splitting of Pd 3d levels (reproduced from ref. 20)

Closer inspection of the spectrum given in Fig. 7.4 shows that emission from some levels (most obviously p, d and f) does not give rise to a single photoemission peak, but a closely spaced doublet. The intensity ratio of the spin orbit split doublet peaks is determined by the multiplicity of the corresponding levels and is equal to $2J + 1$ where J refers the multiplicity factor. The XPS spectrum of Pd sample is shown in Fig. 7.4. Similarly it is seen that $3d_{5/2}$ and $3d_{3/2}$ peaks will have an intensity ratio of 6:4 where as the $4f_{7/2}$ and $4f_{5/2}$ peaks have an intensity ratio 8:6 and the $2p_{3/2}$ and $2p_{1/2}$ will have an intensity ratio 4:2. One can see this aspect in the expanded spectrum for the 3d region of Pd shown in Fig. 7.4.

The $3d$ photoemission is in fact split into two peaks, one at 334.9 eV (BE) and the other at 340.2 eV (BE), with an intensity ratio of 3:2. This arises from spin-orbit coupling effects in the final state. The inner core electronic configuration of the initial state of the Pd is:

$$(1s)^2 \ (2s)^2 \ (2p)^6 \ (3s)^2 \ (3p)^6 (3d)^{10} \ (4s)^2 \ (4p)^6 \ (4d)^{10}$$

with all sub-shells completely full.

The removal of an electron from the 3d sub-shell by photo-ionization leads to a $(3d)^9$ configuration for the final state. Since the d-orbital (l = 2) have non-zero orbital angular momentum, there will be coupling between the unpaired spin and orbital angular momentum. Splitting can be described using individual electron L-S coupling. In this case, the resultant angular momentum arises from the single hole in the d-shell; a d-shell electron (or hole) has l = 2 and s = 1/2, which again gives permitted J-values of 3/2 and 5/2 with the latter being lower in energy.

Table 7.1 Spectroscopic notations used in XPS and also in AES

N	L	J	X-ray level	Electron level
1	0	½	K	1s
2	0	½	L_1	2s
2	1	½	L_2	$2p_{1/2}$
2	1	$3/_2$	L_3	$2p_{3/2}$
3	0	½	M_1	3s
3	1	½	M_2	$3p_{1/2}$
3	1	$3/_2$	M_3	$3p_{3/2}$
3	2	$3/_2$	M_4	$3d_{3/2}$
3	2	$5/_2$	M_5	$3d_{5/2}$
4	3	$5/_2$	N_1	$4f_{5/2}$
4	3	$7/_2$	N_2	$4f_{7/2}$

7.2.2 Multiplet Splitting

Multiplet splitting, also referred to as exchange or electrostatic splitting of core level peaks, can occur when the system has unpaired electrons in the valence levels. For example consider the case of the 3s electron in the Mn^{2+} ion. In the ground state, five 3d electrons are all unpaired and with parallel spins (denoted as ^6S). After the ejection of the 3s electron a further unpaired electron is present. If the spin of the electron is parallel to that of the 3d electron (i.e., the final state ^7S) then exchange interaction can occur, resulting in a lower energy than for the case of anti parallel spin (final state ^5S). Thus the core level will be a doublet and the separation of the peaks is the exchange interaction energy given as E = (2S +1) K3s-3d, where S is the total spin of the unpaired electrons in the valence levels (5/2 in this case) and K3s-3d is the 3s-3d exchange integral. The intensity (I) ratio for the two peaks is given by

$$I (S + ½)/ I(S - ½) = (S +1)/S$$

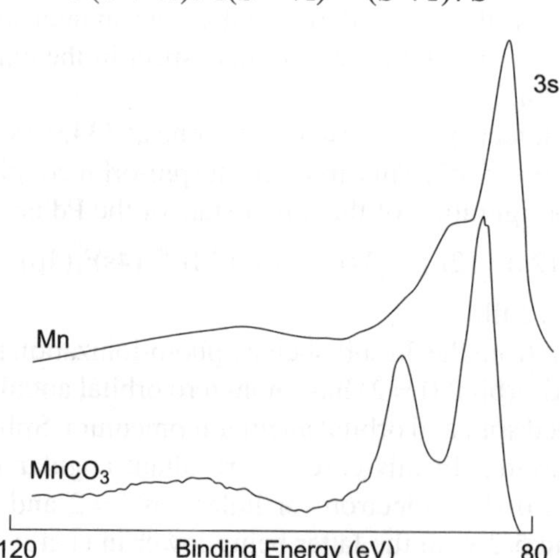

Fig. 7.5 Multiplet splitting of Mn 3s (reproduced from ref. 23)

For the 3s in the Mn^{2+} this ratio is (5/2 + 1) : 5/2 or 1.4:1 and ΔE is calculated to be $\cong$ 13 eV. In the real case of MnF_2, the intensity ratio is 1.8:1 and the separation is $\cong$ 6.5 eV. This discrepancy is due to relaxation and configuration interaction effects, which complicate the simple model described.

The Mn 3s spectrum exhibits a doublet due to multiplet splitting. The energy separation of this doublet depends on the net spin of the Mn atom, providing a direct and element-specific measure of its magnetic moment. However the data for chromium compounds illustrates the fact that this model predicts the trends for changes in the spin state.

Multiplet splitting of the 3s peak from chromium compounds is shown in Fig. 7.6. $Cr(CO)_6$ is diamagnetic (no unpaired electrons); $Cr(C_5H_5)_2$ has two unpaired electrons in the e_g level; and $Cr(hfa)_3$ has three unpaired electrons in the t_{2g} level. It can be seen that the mean BE varies with electronegativity of attached ligands. Multiplet splitting for non s-levels is more complex because of the additional involvement of angular momentum coupling. Thus in Mn^{2+} the 3p level is split into four levels. Multiplet splitting is strongest when both the levels involved are in the same shell (as with 3s, 3p-3d). Multiplet splitting causes broadening (with asymmetry) in both $2p_{1/2}$ and $2p_{3/2}$ peaks leading to apparent variation in the separation which can be used diagnostically. Multiplet splitting is found strongly in the 4s levels of rare earth metals (unpaired electrons in the 4f valence levels) [24].

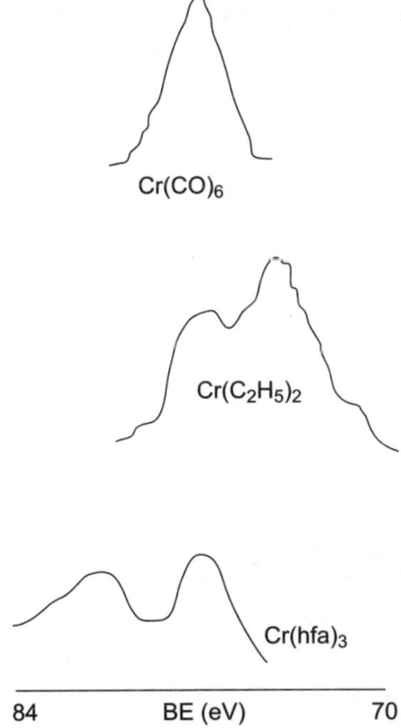

Fig. 7.6 Multiplet splitting in X-ray photoelectron spectrum (reproduced from ref. 24)

7.2.3 Ghost Peaks

Ghost peaks are due to standard X-ray sources which are not monochromatic. Besides the Bremmstrahlung radiation background to the principal $K_{\alpha 1, 2}$ line of magnesium and aluminium targets also produce a series of lower intensity lines, referred to as X-ray satellites. Satellites arise from low probable transitions e.g. K_β, valence band $\rightarrow$ 1s or transitions in a multiply ionized atoms e.g. $K_{\alpha 3, 4}$. X-ray ghost peaks can also be due to excitations arising from impurity elements in the X-ray source. The most common ghost is Al $K_{\alpha 1, 2}$ from Mg K_α source. This arises from the secondary electrons produced inside the source hitting the thin aluminium window. This radiation will therefore produce weak ghost peaks which will appear in the 233 eV higher kinetic energy of those excited by the dominant Mg $K_{\alpha 1, 2}$. Old or damaged targets can give rise to ghost peaks due to Cu L_α radiation the main line from the exposed copper base of the target. These peaks appear at 323.9 eV (556.9) to lower KE of Mg $K_{\alpha 1, 2}$ (Al $K_{\alpha 1, 2}$) excited peaks. In dual anode X-ray sources, which are commonly being used with one anode capable of yielding high energy X-rays (e.g., zirconium, silver and titanium), misalignments inside the source can lead to cross-talk between filaments and anodes, which is an additional source of ghost peaks. X-rays emitted by the copper substrate of aged Mg or Al anodes are often responsible for ghost lines [21]. The spectrum observed is that of the sputtered aluminium sample. The sample was sputtered until the carbon and oxygen adsorbates were removed. The peak at 245 eV binding energy is due to implanted argon. The sample is normally mounted with double sided tape which allowed charging to occur. As a result, the binding energies have shifted approximately by 6 eV. There are ghost peaks present at 323 eV above the aluminium 2s and argon 2p peaks due to photons from the copper mount on the X-ray gun. For a copper source, the most intense peaks have ghost at 323 eV, in the higher binding energy. Similarly a peak at approximately 617 eV binding energy is a ghost resulting from the C (ls) peak at 290.4 eV binding energy [25].

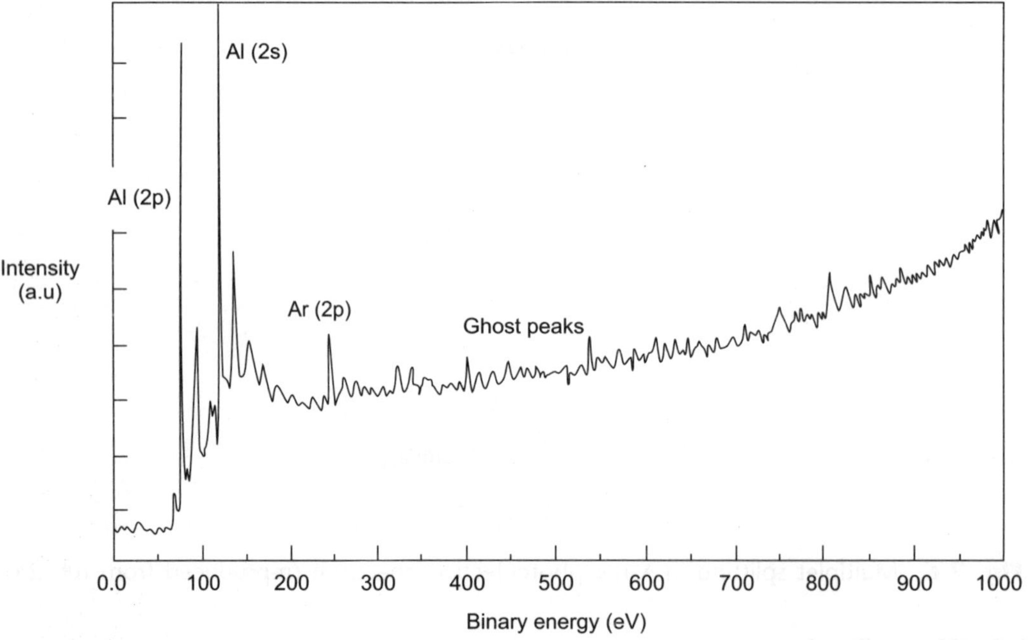

Fig. 7.7 XPS spectrum of a sputtered aluminium sample (ref. 25)

7.2.4 Shake-up and Shake-off satellites

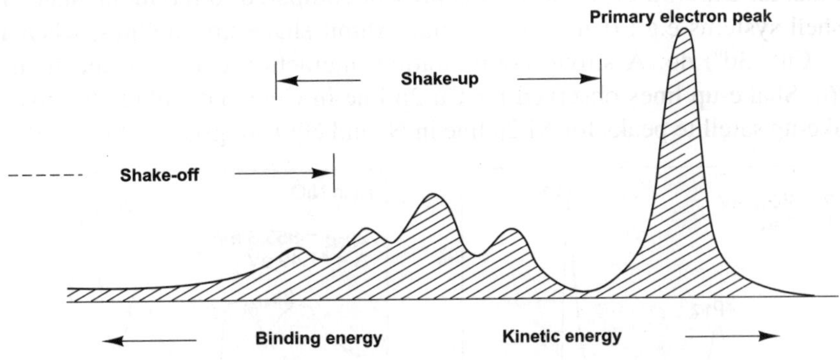

Fig. 7.8 Shake-up and shake off peaks in XPS spectrum

To the valence electrons associated with an atom, the loss of a core electron by photoemission appears to increase the nuclear charge. This perturbation gives rise to substantial reorganization of the valence electrons (referred to as relaxation), which may involve excitation of one of them to a higher unfilled level (shake-up). Fig. 7.8 gives the approximate energy range that both shakeup and shake off will appear in the spectrum. The energy required for the transition is not available to the primary photoelectron and thus the two electron process leads to discrete structure on the low kinetic energy side of the photoelectron peak (shake-up satellites). Strong satellites are observed for certain transition metal and rare earth compounds which have unpaired electrons in 3d or 4f shells respectively. The intensity of satellite peaks is usually higher than what would be expected for atomic–like shake-up and the reason for this has led to a great discussion.

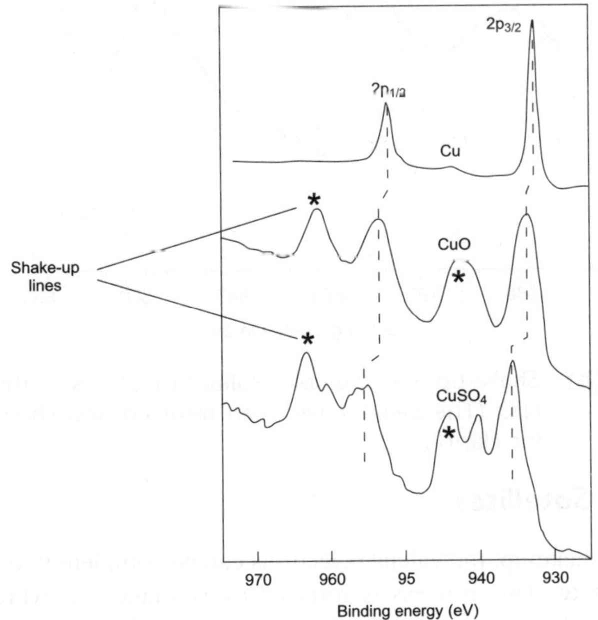

Fig. 7.9 Examples of shake-up lines (*) of the copper 2p observed in copper compounds (reproduced from ref. 23)

The convincing explanation is that in the final state there is significant ligand to metal charge transfer such that an extra 3d or 4f electron is present compared to the initial state. This explains why closed shell systems e.g.: (Cu^{2+} $3d^{10}$) do not exhibit shake-up satellites, whereas open shell systems e.g., (Cu^{+}, $3d^{9}$) do. A strong configuration interaction occurs in the final state due to relaxation [26]. Shake-up lines observed for Cu 2p line in CuO and $CuSO_4$ are given in Fig. 7.9. Similarly shake-up satellite peaks for Ni 2p line in Ni and NiO are given in Fig. 7.10.

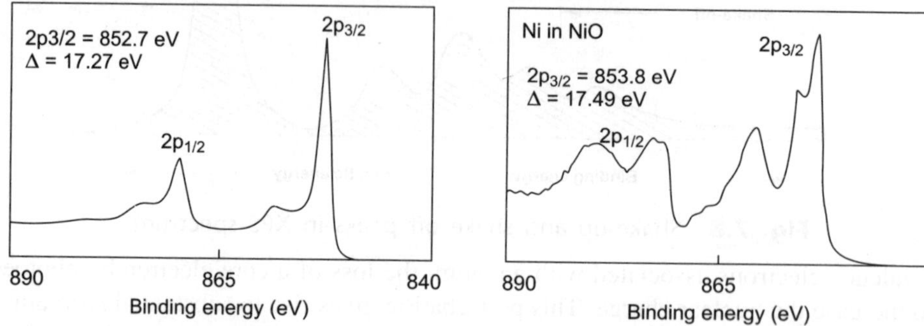

Fig. 7.10(a)   Ni and NiO shake-up peaks (reproduced from ref. 23)

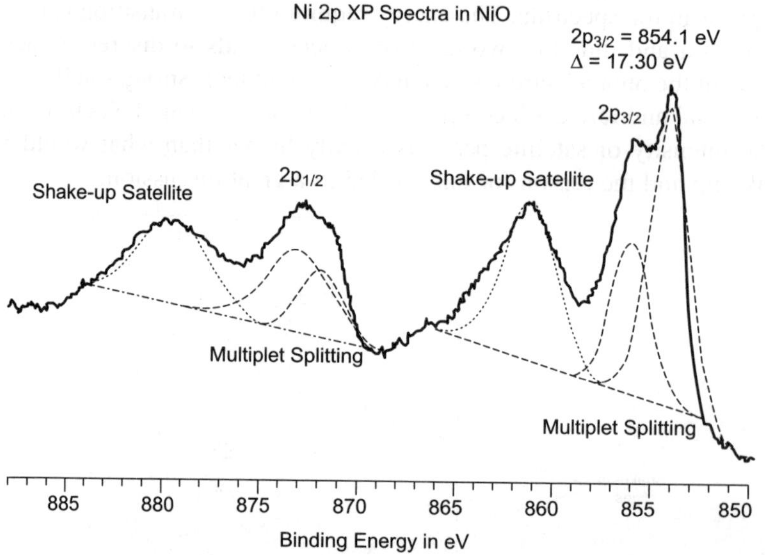

Fig. 7.10(b)   Shake-up and multiplet splitting features in the Ni 2p region in NiO (The peake have been resolved and shown by dashed lines for clarity)

7.2.4.1 Shake-off Satellites

In a process similar to shake-up, the valence electrons can be completely ionized i.e., excited to an unbound continuum state. This process is referred to as shake off, which leaves an ion with vacancies in both the core level and the valence level. Discrete shake off satellites are rarely discerned in the solid state because:

1. The energy separation from the primary photoelectron peaks are greater than for shake–up satellites, which means the satellites tend to fall within the region of the broad inelastic tail.
2. Transitions from discrete levels to a continuum produce onsets of increased intensity (i.e., broad shoulders) rather than discrete peaks.

Normally XPS is considered as a nondestructive technique. However, in some cases this may not hold good. Ultra high vacuum may produce a loss of water from the sample, which is a main issue in the case of biological systems. Radiation damages in the polymer systems and the Cu-S systems are also serious issues.

7.2.5 Asymmetry in Core Level Peaks

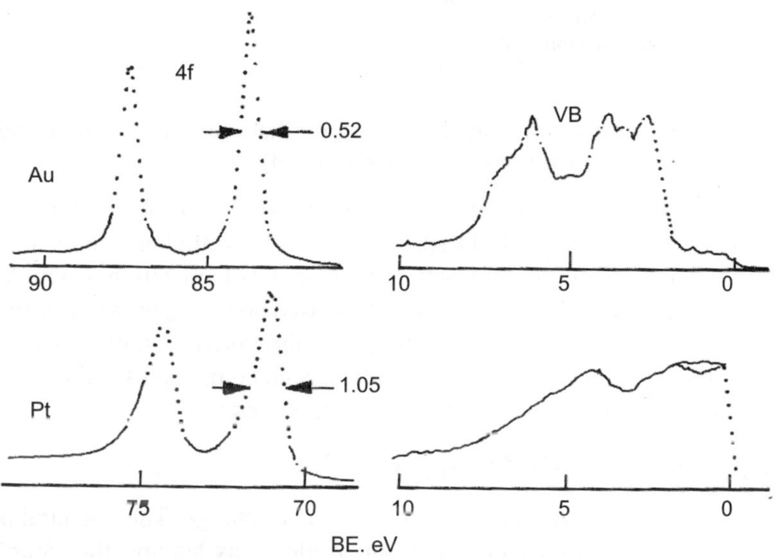

Fig. 7.11 Core (4f) and valence (VB) photoelectron spectra of gold and platinum recorded using monochromatic Al K_α radiation (ref. 24)

7.2.5.1 Asymmetry of Core Levels in Non-metals

In a solid metal there exists a distribution of unfilled one-electron levels above the Fermi energy, which are available for shake up type events following core electron emission. In this case, instead of discrete satellites to low kinetic energy of the primary peak, the expectation is for the core level itself to be accompanied by a low kinetic energy tail. The higher the density of states at the Fermi level, the more likely is this effect to be observed. The relationship between the degree of core level asymmetry and the density of states can be seen from Fig. 7.11.

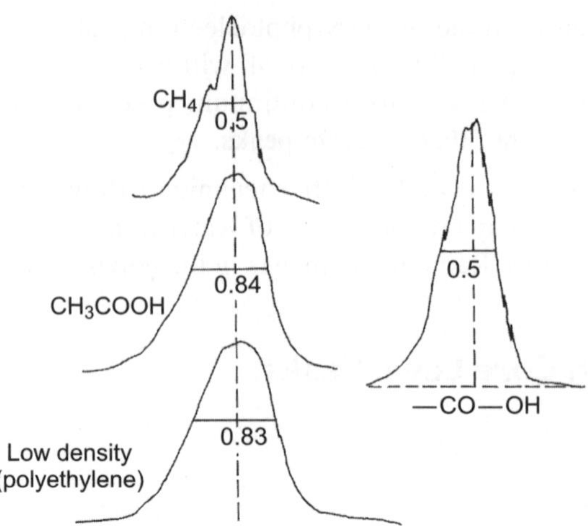

Fig. 7.12 C 1s core level spectra from gas phase methane and acetic acid and solid state insulating low density poly ethylene (ref. 24)

This effect, which is generally observed in the recorded spectra, obtained at the very highest energy resolution is illustrated by the C 1s spectrum from polyethylene (refer to Fig. 7.12). The marked asymmetry is due to vibrational fine structure. Core level ionization in a molecule leads to a redistribution of electronic charge and in general to decreased bond lengths and a narrower molecular potential curve. Thus core ionization is likely to result in vibrational excitation and a core level peak composed, in the harmonic approximation, of equally spaced components (vibrational energy level spacing) with decreasing intensity towards higher binding energy.

7.2.6 Charging and Sample Damage

Samples ejecting photoelectrons will lead to accumulation of charge. The potential acquired by the sample is determined by the photoelectric current of electrons leaving the sample, the current through the sample holder towards the sample and the flow of the Auger and secondary electrons from the source window onto the sample. Due to the accumulation of the positive charge on the sample, all XPS peaks in the spectrum shifts to the higher binding energies. Normally calibration is done using the binding energy of a known compound for example Si 2p in SiO_2 is taken as 103.4 eV or carbon 1s binding energy is considered as 284.4 eV. There are other substances which can be added and also used for calibration. In addition the shifting, peaks can also broaden and thus decrease the resolution with a low signal to noise ratio. Charging in supported catalysts is usually limited to a few eV in normal mode of operation but when monochromatic XPS is used it can be considerably higher. These effects can be minimized by using a flood gun which showers the low energy electrons on the sample or mounting the sample like powders in the indium foils.

Two additional approaches to charge compensation have emerged recently. One is the dual-beam charge neutralization system, which includes a low-energy electron beam and a low-energy ion beam. The ion beam's job is to neutralize any electrostatic charges on the sample that might repel the electron beam from the analysis area. The second approach is to place a magnetic immersion lens and a source of low-energy electrons at the entrance to the energy analyzer. The magnetic field traps the low-energy electrons between the bottom of the lens and the sample surface. Low-energy electrons are consumed until the charge on the surface is neutralized; at which point the remaining low-energy electrons are repelled and simply oscillate in the magnetic field [24].

7.2.7 Surface Plasmon Loss Features

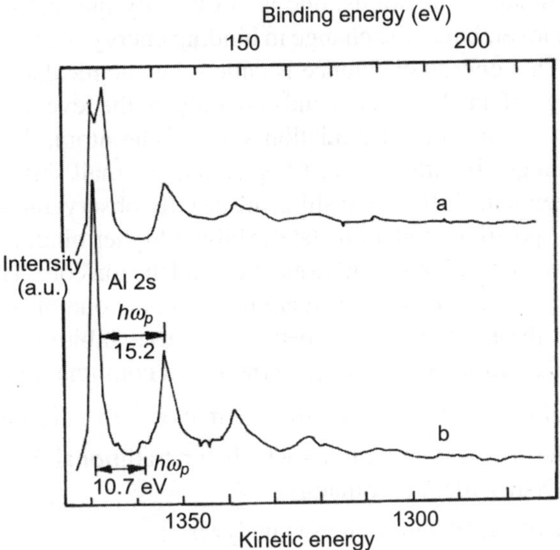

Fig. 7.13 Surface plasmon loss spectra of Al 2s , (a) Slightly oxidized aluminium, (b) Clean aluminium (ref. 27)

Surface plasmon loss peak arise when outgoing electrons of sufficient energy, originating in any way, lose energy by exciting collective oscillations in the conduction electrons in a solid. The oscillation frequency and hence the energy involved, is characteristic of the nature of the solid and is called the plasmon frequency. The plasmon loss determined by three dimensional nature of the solid is called a 'bulk' or 'volume' loss, but when it is associated with the two dimensional nature of the surface is called a 'surface' loss. Both types of plasmon losses from the Al 2s peak can be seen in Fig. 7.13 with a bulk loss of 15.2 eV and a surface loss of 10.7 eV. Theoretically the surface plasmon loss energy should be a factor of $2^{1/2}$ less than the energy of bulk plasmon losses. When there is a pronounced bulk loss as in a free electron – like metal such as aluminium, a succession of peaks can be found at multiples of the loss energy with progressively decreasing intensity as shown in Fig. 7.13. Spectrum (a) from slightly oxidized aluminium reveals the rapid quenching of the surface plasmon on oxidation and also shows a chemical shift in the Al 2s photoelectron peak. In the XPS spectrum from a solid a series of plasmon loss peaks is associated with prominent peaks. The fundamental or first plasmon loss is always discernable. Depending on the material and experimental conditions, several multiple plasmon losses of decreasing intensity may also become visible as can be seen from this example.

7.3 IMPORTANT FEATURES OF THE X-RAY PHOTO ELECTRON SPECTROSCOPY

7.3.1 Chemical Shifts

In the investigation of molecules and solids, one is not usually interested in the absolute binding energy of a particular core level, but in the change in binding energy between two different chemical forms of the same atom. This energy difference is called the chemical shift.

The exact binding energy of an electron depends not only on the level from which photoemission is occurring, but also on: 1. The formal oxidation state of the atom, 2. The local chemical and physical environment. Changes in either (1) or (2) give rise to small shifts in the peak positions in the spectrum, so-called chemical shifts. Such shifts are readily observable and interpretable in XPS spectra. Atoms of a higher positive oxidation state exhibit a higher binding energy due to the extra coulombic interaction between the photo-emitted electron and the ion core. This ability to discriminate between different oxidation states and chemical environments is one of the major strengths of the XPS technique. In practice, the ability to resolve between atoms exhibiting slightly different chemical shifts is limited by the peak widths which are governed by a combination of factors, especially

1. The intrinsic width of the initial level and the lifetime of the final state.

2. The line-width of the incident radiation—which for traditional X-ray sources can only be improved by using X-ray monochromators.

3. The resolving power of the electron-energy analyzer.

In most cases, the second factor is the major contribution to the overall line width.

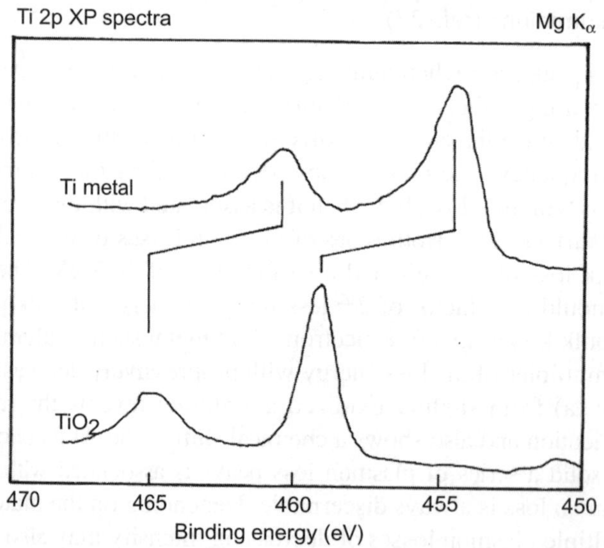

Fig. 7.14 Oxidation States of titanium (ref. 20)

The two spin orbit components exhibit the same chemical shift. Titanium exhibits very large chemical shifts between different oxidation states of the metal; Ti 2p spectrum from the pure metal (Ti) is compared with a spectrum of titanium dioxide (TiO_2) in Fig. 7.14.

7.3.2 Small Area Resolution

Spectroscopy on small-area works essentially the same way, except that the XPS instrument zooms in on a few micrometer area of the sample. Decrease of the analysis area is made possible either by focusing the X-ray source to different spot sizes, so-called microprobe instruments or by restricting the analysis area with a physical aperture and an extra set of lenses in front of the energy analyzer. Whereas a minimum analysis area of 20-30 μm^2 was common five years ago, some current instruments can zoom in on an area of 10-15 μm^2.

Photo electron spectrum is a plot of the intensity of the photoelectrons as a function of electron kinetic energy. In photoelectron spectroscopy the electron kinetic energy that one normally encounters is of the order of 10 – 1000 eV. For electrons of this energy, the distance traveled before an inelastic encounter is of the order of 20-30 Å. This distance corresponds to about 10 atomic layers of the material, in most cases. Therefore, while analyzing kinetic energy of electrons, the electrons of maximum intensity which do not suffer energy loss due to inelastic collisions are principally coming from the very top of the surface and therefore, the technique is extremely surface sensitive. The distance in which the inelastic collisions do not occur is called the inelastic mean free path and is a quantity very difficult to measure. Normally what can be measured is attenuation length, which is a combination of both inelastic and elastic mean free path. For metals the attenuation length is of the order of 20 Å.

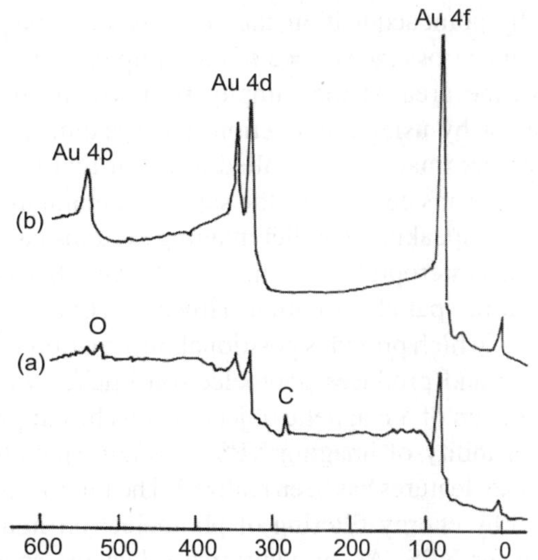

Fig. 7.15 Spectrum of a gold sample (a) before and (b) after Ar ion bombardment Ar ion bombardment is used generally for cleaning the surface

7.3.3 Depth Profile

Since XPS instruments have small-area analysis capability and improved sensitivity, they are more analogous to Auger instruments in terms of depth profiling. The universal method of destructive composition-depth profiling applied in AES and XPS is through the sputtering of thin sections of the sample surface by energetic ion bombardment. It has been done using a small analysis ion beam which can be focused on the sample and can etch the surface quickly. Compared to AES, depth profiling XPS is more quantitative, provides more chemical information, and has less difficulty with insulating films. A typical example is shown in Fig. 7.15. It can be seen that the intensity of emissions from the energy levels of gold (namely 4f, 4d, 4p) is increased on Ar ion bombardment.

7.3.4 Imaging XPS (iXPS)

Another area where XPS instruments have improved significantly, according to the experts, is photoelectron imaging, which maps a sample surface. Images can be acquired in several ways. In point-by-point acquisition, the image is built one pixel at a time. One way to do this is to scan the X-ray beam across the surface of the sample. The other option is to keep the X-ray beam in place and change the area on the sample from which photoelectrons are extracted, either by moving the sample or by using a deflection lens system. Instead of point-by-point acquisition, some systems use parallel imaging also called direct or real-time imaging in which the spatial relationship of the photoelectrons coming off the sample is maintained, and the entire image field is collected at once. Generally speaking, parallel imaging systems can achieve a spatial resolution of ~ 2-3 µm, whereas a scanned microprobe system can achieve ~ 8-10 µm [21]. The inherent weakness in XPS has been the lack of spatial resolution. However some modern instruments offer the capability of parallel imaging, which provides positional information from dispersion characteristics of the hemispherical analyzer and produces photoelectron images with spatial resolution better than 5 mm. Also, the introduction of a magnetic objective lens has improved the sensitivity for a given spatial resolution. The capability of imaging XPS to obtain spatially-resolved chemical maps across the surface of very small features has been realized. The microscope allows imaging with chemical contrast (Imaging ESCA) by energy filtering of photoelectron images at kinetic energies upto 1.6 KeV, which are typical for XPS. As an additional effect the spatial resolution of the images is enhanced by the reduction of the chromatic aberrations, even when using threshold photoemission.

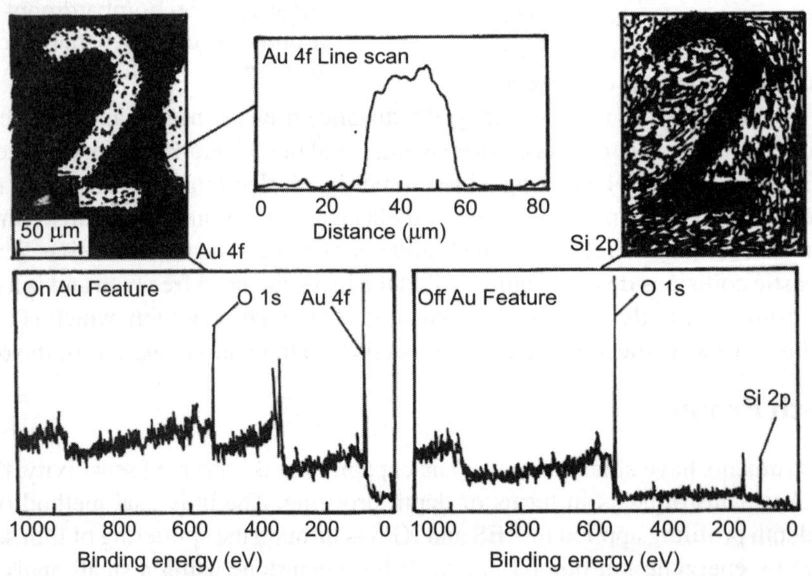

Fig. 7.16 Parallel XPS images from gold features on a glass substrate (ref. 22)

The resolution in the Au 4f and Si 2p images can be measured from the line scan to be about 3 µm. The images can be used to define small areas from which spectra can be acquired. In Fig. 7.16, XPS image of Au feature on a glass substrate is given.

7.3.5 Angular Effects

Photo electron spectra can be strongly influenced by the particular geometry employed in their measurement, i.e., the relative orientation of the source, sample and spectrometer (lens). Two types of angular effects are important. The first involves the increase in surface sensitivity obtained at low angles of electron exit relative to the sample surface (take off angle) and is particularly important in XPS. The second is concerned with the single crystal materials.

Surface Sensitivity Enhancements by Variation of the Electron Take-off Angle

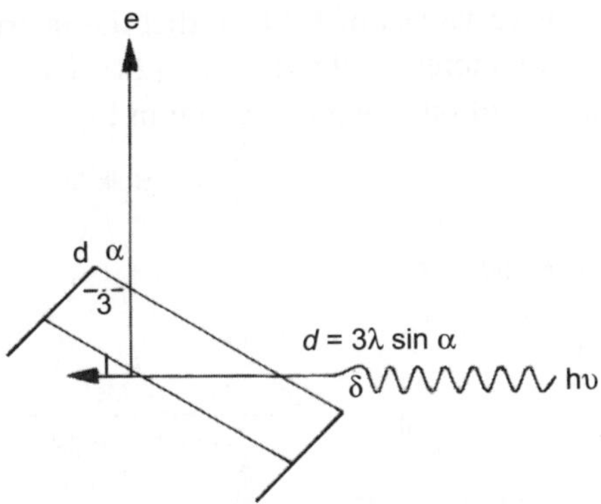

Fig. 7.17 Surface sensitivity enhancement by variation of electron take-off angle (ref. 24)

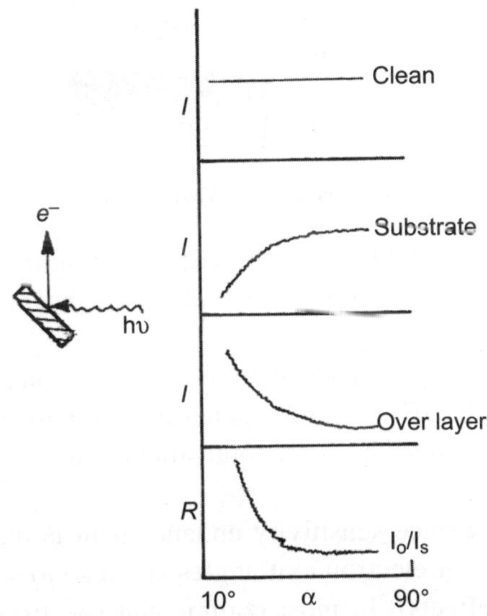

Fig. 7.18 Theoretical angular dependent curves for a clean flat surface and a flatover layer/substrate surface (ref. 24)

If λ is the attenuation length (AL) of the emerging electron then 95% of the signal intensity is derived from a distance 3λ within the solid. The vertical depth sampled is given by $d = 3\lambda \sin\alpha$

The angular variation of intensities is given by

$$I_s^d = I_s \exp\left(-d/\lambda \sin\alpha\right)$$

$$I_o^d = I_o \exp\left(1-\exp\left(1 - e\alpha p\left(-d/\lambda \sin\alpha\right)\right)\right)$$

Where λ is the appropriate value of attenuation length for the observed photoelectron, α is the photoelectron take off angle, d is the vertical length sampled and I_s^d, I_o^d are the intensity of photoemission from substrate and over layer respectively. In the ideal situation these equations lead to curves of the type as shown in Fig. 7.18. However the real case is actually complicated by the fact that the system geometry imposes a response function also dependent on α. This complication is avoided by measurements of relative values of I_o / I_s so that the instrument response function is cancelled. Thus at low values of α, I_o / I_s increases significantly [24]. The change in the peak intensities when measured with a surface angle and bulk angle is given in Fig. 7.19.

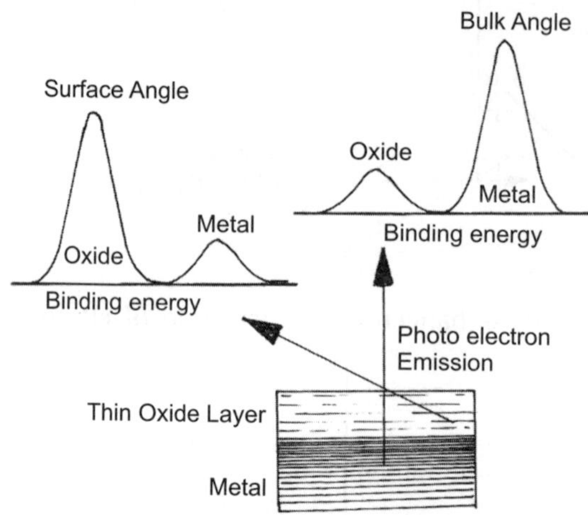

Fig. 7.19 Illustration of XPS spectra taken from a thin oxide film (ref. 22)

Effect of variation of the take –off angle on the Si 2p spectrum from silicon with a passive oxide layer is shown in Fig. 7.20. It can be seen that the relative enhancement of the surface oxide signal is taking place at low angles.

A series of Si 2p photoelectron spectra are recorded for emission angles of 10-90° to the surface plane. The Si 2p peak of the oxide (BE ~ 103 eV) increases markedly in intensity at take off emission angles whilst the peak from the underlying elemental silicon (BE ~ 99 eV) dominates the spectrum at near-normal emission angles.

The major requirement for surface sensitivity enhancement is that the surface is flat. Surface roughness leads to an averaging of electron exit angles and also to shadowing effects (both of the incident X-ray and emerging effects). In most commercial spectrometers the angle between the incident X-rays and detected electrons (180-α-δ) is fixed. At very low delta there is also a surface sensitive enhancement effect due to rapid fall off in X-ray penetration depth as δ tends to grazing incidence. AES instruments are generally operated with large input apertures to maximize sensitivity, with consequent loss of angular resolution.

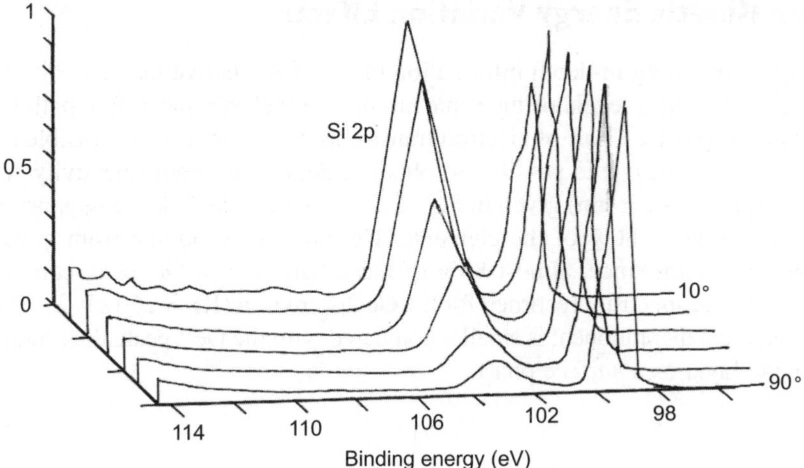

Fig. 7.20   Angle-dependent analysis of a silicon wafer with anative oxide surface layer (ref. 20)

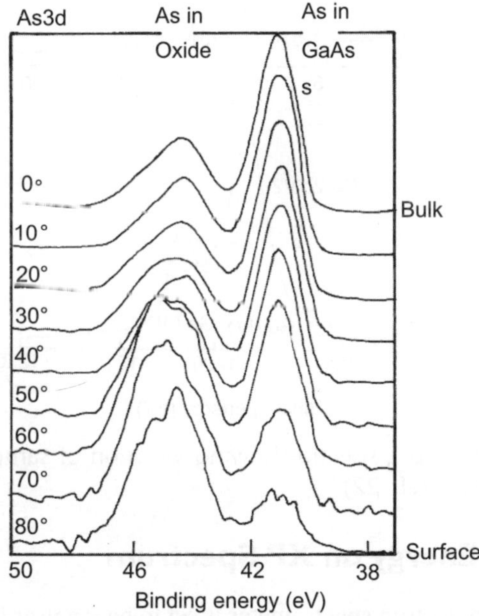

Fig. 7.21   Angle resolved XPS data acquired by tilting the specimen. In this case the specimen is gallium arsenide with a thin layer of oxide at the surface. (ref. 22)

It is clear from Fig. 7.21 that the spectra of As 3d region is dominated by the oxide at the surface and falls off as the probing depth increases. This phenomenon is also repeated in the gallium part of the spectrum. Thus straightway one can conclude that the bulk GaAs is covered by a layer of material in which Ga and As are oxidized. It is possible to determine the over layer thickness and over layer coverage from angle resolved XPS studies.

7.3.6 Electron Kinetic Energy Variation Effects

An alternative way of obtaining in-depth information in a nondestructive manner is by examining electrons from different energy levels of the same atom. The inelastic mean free path varies with kinetic energy, and by selecting a pair of electron transitions which are both accessible in XPS but have widely separated energies; it is possible to obtain a degree of depth selectivity. The Ge 3d spectrum (KE = 1450 eV, $\lambda \sim 2.8$ nm) given in Fig. 7.22 shows Ge° and Ge^{4+} components with the oxide component being about 80% of the elemental Ge. i.e., in Ge 3d spectrum given Ge (IV) contribution is the maximum since, almost 80% of the intensity of a Ge° peak comes from the Ge (IV) only. Similarly the spectrum recorded for the Ge $2p_{3/2}$ region (KE = 260 eV, $\lambda \sim 0.8$ nm) one can notice that the elemental component is smaller compared with the Ge^{4+} peak, thus confirming the presence of the oxide layer as a surface phase.

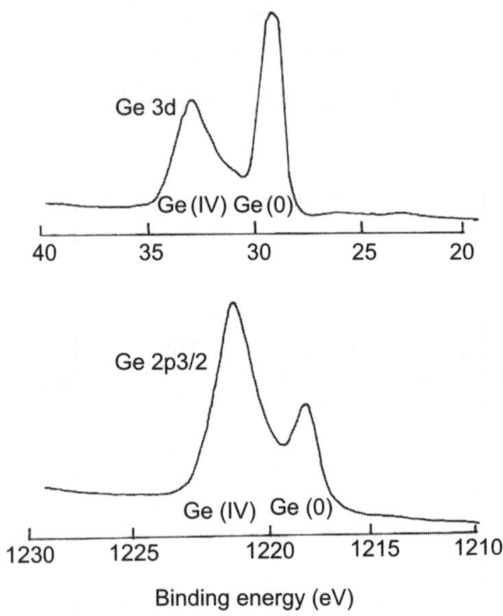

Fig. 7.22 Ge 3d and Ge $2p_{3/2}$ spectra showing variation of sampling depth with electron kinetic energy (ref. 22)

7.3.7 Effect of Pass Energy on XP Spectrum

The analyzer in X ray photoelectron spectrometer is set to pass a user defined fixed electron energy to the electrons ejected. The selected pass energy affects both the transmission of the analyzer and its resolution.

Selecting a low pass energy will result in high resolution whereas a high pass energy will provide higher transmission but poor resolution. As the pass energy remains constant throughout the electron kinetic energy range, the resolution (in electron volts) is constant across the entire width of the spectrum.

The XPS spectrum of silver recorded at a series of pass energies is shown in Fig. 7.23. The effect of pass energy on resolution can be seen. The spectra have been normalized, but the increase in noise with decrease in pass energy is a result of the decrease in sensitivity. In a typical XPS experiment, the user will select a pass energy in the region greater than 100 eV for survey or wide scans and in the region of 20 eV for higher resolution spectra of individual core levels. These narrow scans are used to establish the chemical states of the elements present and for evaluating quantitative data. It is normal practice to collect XPS spectra in the constant analyzer energy mode, so that the energy resolution (in electron volts) remains constant across the spectrum.

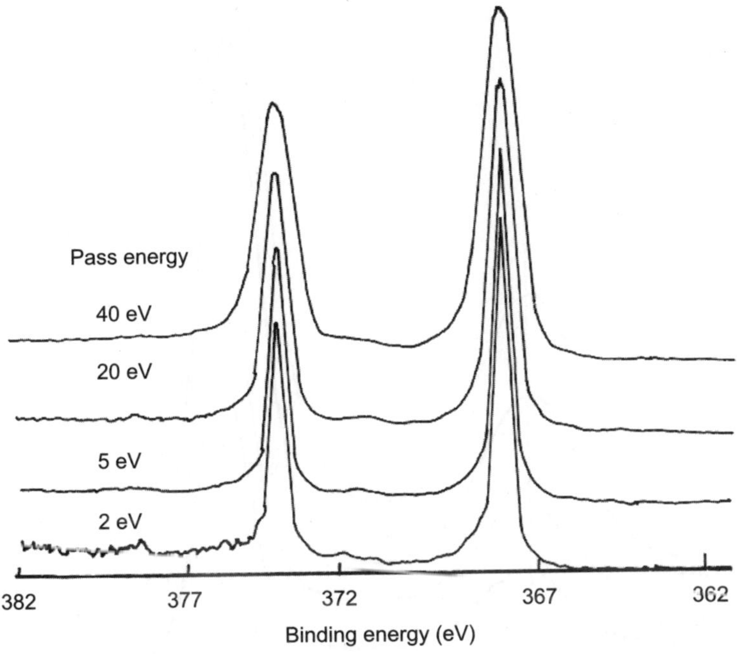

Fig. 7.23(a) XPS spectrum of silver showing the effect of pass energy upon the Ag 3d part of the spectrum (ref.27)

7.4 XPS IN CATALYSIS

7.4.1 Applications of XPS for the Study of Catalysts

In fundamental catalytic research, it is often attractive to study the behaviour of non- supported catalysts in order to exclude the direct or indirect influence of the support on the catalyst performance. For XPS examination, such non-supported catalyst powders are attractive too, because different properties can be studied separately and the XPS signals are relatively strong. But a working catalyst is not as simple and the study of working catalyst system is a complex phenomenon in which a multitude of interesting variables intervene. The dispersion of metal catalyst on a nonmetallic porous

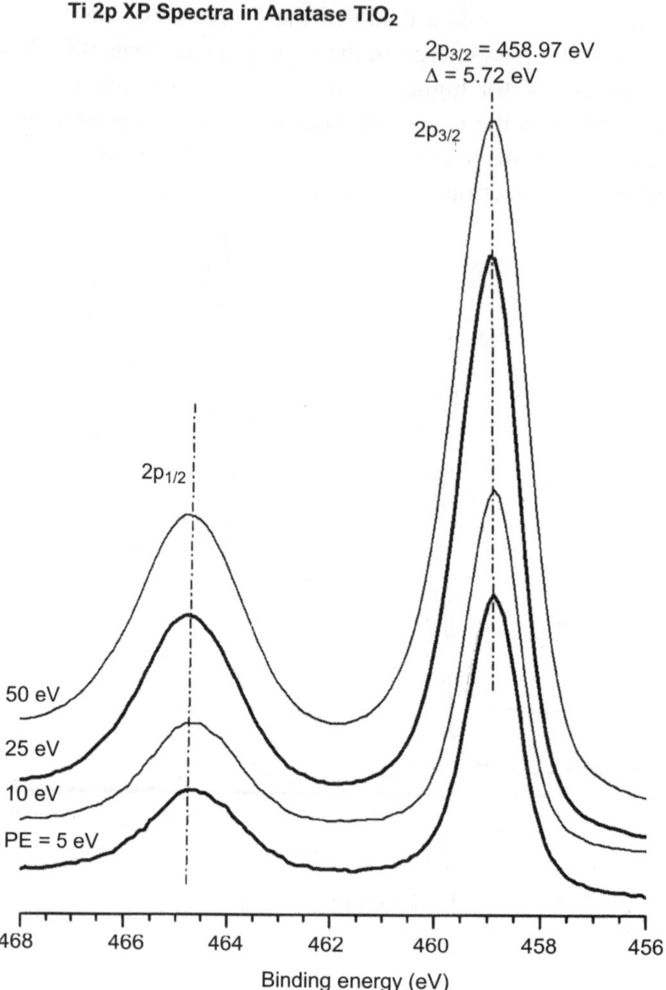

Fig. 7.23(b)   Ti 2p X-ray photoemission region for a sample of anatase TiO2 using different pass energies

medium has long been used in practical catalysis. It is convenient for obtaining a relatively large area which is resistant to sintering during its use, or regeneration. Supports are believed to be inert materials acting only to disperse physically the active component. It is rather difficult to find a totally inert support and the interaction of the support with the metal influences the properties of the catalysts. The physico-chemical characteristics of the support also influence the distribution of the active metal and in turn, the catalytic activity. Other factors which influence the properties and activities of the catalysts are the preparation procedures and the pre-treatment conditions. Apart from this, the catalysts can undergo changes during the reaction. Hence an understanding of the catalyst from the preparation stage to spent stage helps in designing new catalysts with improved stability, activity and selectivity.

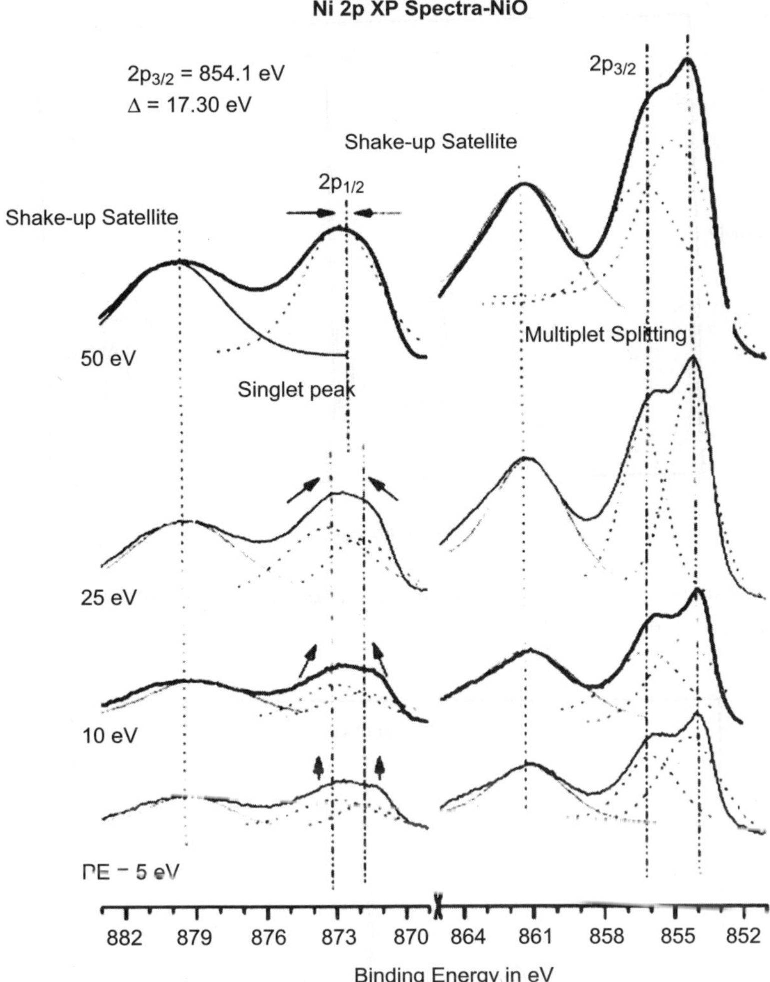

Fig. 7.23(c)   Illustration of the disappearance of the multiplet fratures in Ni 2p XP emission region with increasing energy

7.4.1.1 *XPS Analysis of Supports and Catalysts*

Impurities in the support can influence their interaction and also the catalytic activity. It is, therefore, desirable to identify the impurities by XPS analysis of the support as well as of catalyst. Wide scan XPS spectra of two different support samples are shown in Figs. 7.24 and 7.25. As can be seen from the spectrum, (Fig. 7.24) the sample is extremely pure as no impurity or carbon was detected. However the other sample (Fig. 7.25) differs from the first one and contaminating elements such as Na, Ti, Ca, C, Cl and Si were detected.

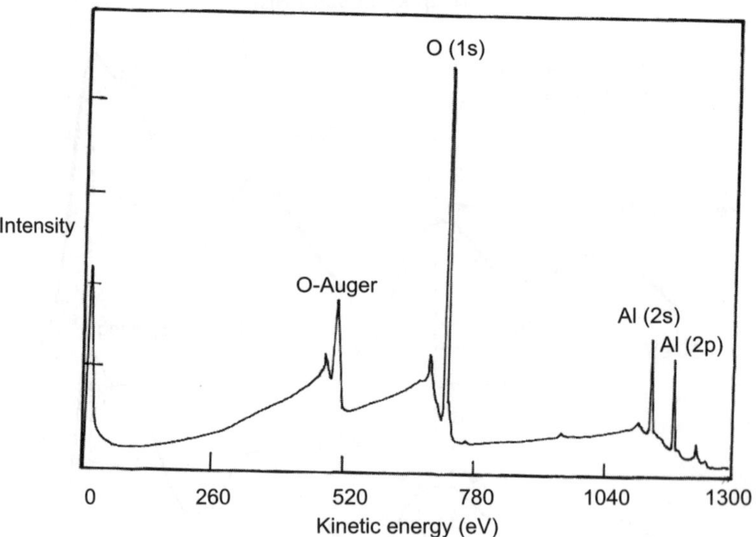

Fig. 7.24 Wide scan XPS spectrum of an alumina support material showing an extremely pure surface. (ref. 29)

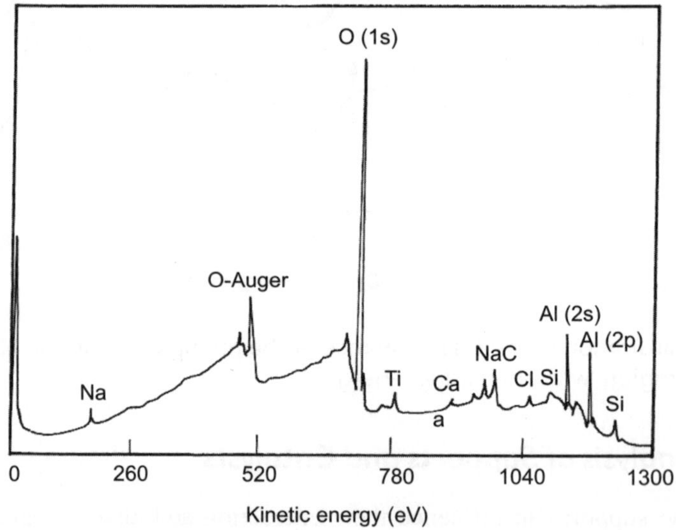

Fig. 7.25 Wide scan XPS spectrum of a contaminated alumina support material (Ref. 29)

7.4.1.2 *Dispersion in Supported Metal Catalyst*

The activity of supported metal catalyst is frequently determined by the degree of dispersion of the active species which have been deposited on the carrier.

Dispersion D may also be defined as:

(Number of surface metal atoms per unit weight of catalyst)

Total number of metal atoms per unit weight of the catalyst

Dispersion in some cases (e.g., metals) can be directly monitored by selective adsorption of suitable gases. In other cases, e.g., oxides or sulphides, where no such selective adsorption is known, indirect physico-chemical methods have to be employed. XPS intensity measurements have been used in this connection and several authors have developed the theoretical basis of the quantitative interpretation of the data. Care must be exercised in the interpretation of XPS intensity measurements, applied to the study of dispersion of catalysts. Inhomogenisation and repartition of active species affect the intensity measurements. For an accurate interpretation, a structural model of the surface layer is required. For a homogeneous solid with a flat surface, the following relation has been proposed:

$$I = k\,\sigma\,n\,\lambda\,F$$

where k is a spectrometer constant; σ, the photo electron cross section for a given solid; n, the concentration of atoms undergoing photo electron ejection; F, the X-ray flux; and the λ, the electron mean free path for inelastic collision. In the case of porous materials such as supported metal catalysts, the model leading to the integrated intensity should take into account metal crystallite size distribution of metal within the pore structure and the effect of surface roughness. Some simplified models for the quantitative analysis of supported catalysts have been proposed. Many of the various factors involved in the expression of the integrated intensities cancel out when the ratios of the intensities of the two key components are compared. For a uniform monoatomic dispersed material, the ratio $I_{metal}/I_{support}$ derived from the integrated areas corrected for cross section should be proportional to the ratio of the number of metal atoms per unit area of the support. If the intensity ratio is higher than expected from the metal surface density, it indicates migration of the metal to the external surface. If it is lower it shows particle growth or preferential deposition in small pores not easily seen from the exterior of the particle. In the case of bimetallic catalysts, the integrated intensity ratios, corrected for photoelectron cross sections, can be compared with atomic ratios of the metals as loaded in the catalysts. Deviation from initial stoichiometry would suggest surface enrichment of one metal in preference to the other. Typical plots of ESCA intensity ratios vs. metal loading for heterogeneous catalysts are shown in the Fig. 7.26. I_m/I_s is the ESCA intensity ratio of supported metal (I_m) to support (I_s). In the Fig. 7.26 (A) represents high surface area support with considerable dispersion, (B) represents medium surface area with changing dispersion and (C) represents medium surface area with pore filling and repartition.

The plot of Ni/Al ESCA intensity vs. their bulk metal content for a series of nickel containing catalysts is shown in Fig. 7.27. A linear variation is observed between 0 and 17 wt% of Ni, but shows a deviation around 17 wt% and the slope increases. In 0-17 wt% range, nickel binds to the alumina lattice in a highly dispersed fashion. At percentages higher than 17 wt%, crystallites of NiO begin to form on the surface. Hence, the intensity variation with composition shows two linear regions. This analysis has enabled to detect the change in the nature of the Ni species present on the catalyst.

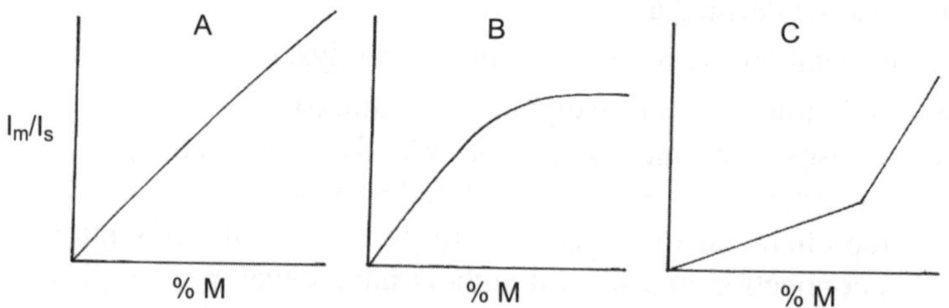

Fig. 7.26 Typical plots of ESCA intensity ratios vs. metal loading for heterogeneous catalysts, (A) high surface area; constant dispersion; (B) medium surface area, changing dispersion; (C) medium surface area, pore filling and repartition. I_m/I_s = ESCA intensity ratio of supported metal (I_m) to support (I_s) (ref. 29)

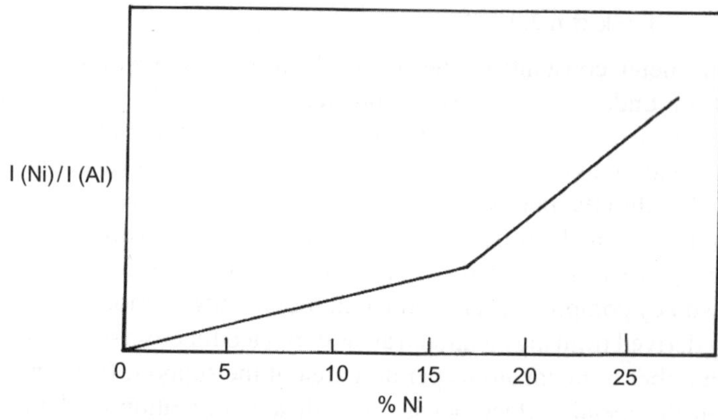

Fig. 7.27 ESCA intensity plot for Nickel on alumina catalyst of varying nickel content (ref. 30)

Consider a catalyst system where crystallite growth of metal particles takes place not just by agglomeration but due to sintering at high temperature treatment. A correlation can be expected between I_m/I_s intensity ratio (by XPS) and the metal crystallite size measured by XRD. Measurements were made on catalysts containing the same percentage of metal but treated at different temperatures to produce metal crystallites of different sizes. As the metal particles become larger, I_m/I_s ratio measured by ESCA decreases. This is consistent with interpretation that more of the metal dispersed over the surface in small particles can be seen by the spectrometer than for large particles. Not all photoelectrons ejected from metal atoms can escape from the large particles; where as most of the electrons can escape from the small particles.

7.4.1.3 ESCA Intensity Ratio as a Probe of the Catalyst Structure

One of the unknowns in the studies of heterogeneous catalysis is the manner in which the catalytically active component interacts with the physical structure of the support material. Since ESCA is a tool for the surface characterization, the intensities of the observed electron lines reflect surface structure of the supported metals and oxides. Information on the depth of metal (Rh) penetration into the pores of the carbon supports of different physical properties was obtained from ESCA measurements on Rh/C catalysts by Brinen and Schmitt [31].

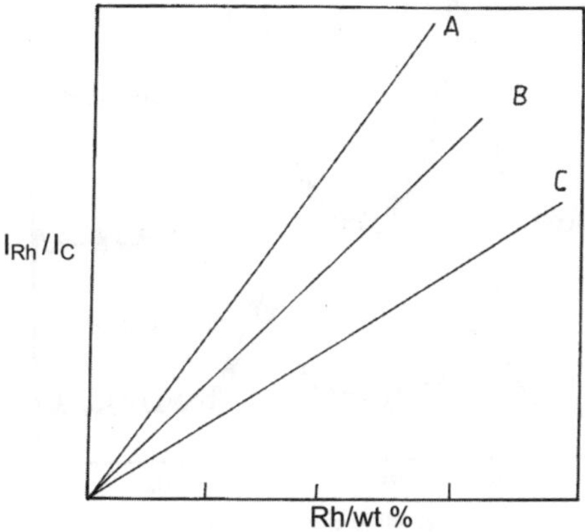

Fig. 7.28 Variation of I_{Rh} / I_C for various loadings of rhodium on three different carbons. carbon A = 10 m²/g; carbon B = 850 m²/g with average pore radius 11 Å; carbon C = 560 m²/g with average pore radius 53 Å (ref. 29)

Intensity ratios vary directly with metal loading for three catalyst supports A, B and C. However intensity ratios are different for different carbon supports and cannot be explained on the basis of surface area alone, since large surface area carbon B (850 m²/ g) has intensity ratio (I_{Rh}/I_C) intermediate between carbon A (10 m²/g) and carbon C (560 m²/g). The difference can be attributed to the pore structure, since surface area of carbon A is low, Rh on impregnation can form 5-6 atomic layers on the support and a large fraction of Rh is accessible to ESCA. Carbon B is microporous and Rh is preferentially deposited in the outer portion of the pores during catalyst preparation and is expected to be non-uniform. The effect is less important for carbon C since Rh particles are easily coated on the large pores and the intensity variation is uniform throughout the support and may also represent high dispersion. These results suggest that the understanding of the physical structure of the support is required for the proper interpretation of the ESCA spectra of metals loaded on porous supports. In the case of reactions such as hydrogenation of benzene, which is a facile reaction, the activity is directly dependent on the accessible metal surface, and hence a direct correlation between the catalytic activity and intensity ratio (I_m/I_s) is possible.

7.4.1.4 *Effect of preparation methods*

Au supported on CeO_2 is a good catalytic system for the oxidation of the volatile organic compounds. When two different methods such as deposition precipitation (DP) and precipitation deposition (PD) are used for the preparation of Au/CeO_2, both showed different activities. X-ray photo electron spectra of these two samples are given in Fig. 7.29.

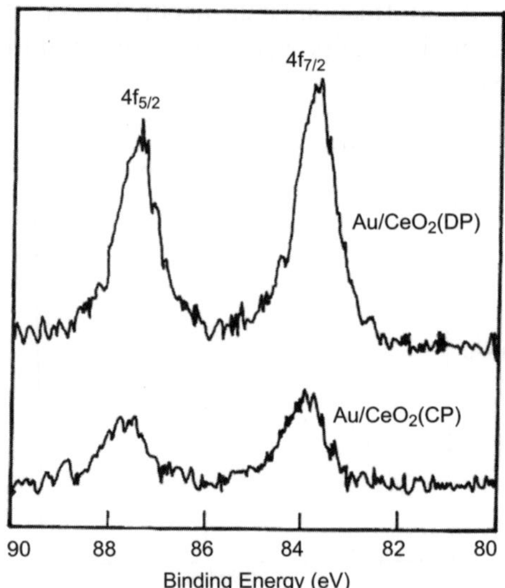

Fig. 7.29 XPS spectra of Au/CeO$_2$-prepared by coprecipitation (CP) and deposition -precipitation (DP) (ref. 32, 33)

It can be seen that the metallic gold on the ceria support is more if the preparation method is deposition precipitation (DP). So the Au/CeO$_2$ (DP) will exhibit better catalytic activity since higher amounts of active species is present on the surface.

7.4.1.5 Correlation between the Atomic Ratios and the Activities

A highly dispersed Ta in a mesoporous like matrix was used for the liquid phase sulfoxidation of 4, 6-dimethyl-2-thiomethylpyrimidine to the corresponding sulfoxide. Ta$_2$O$_5$–SiO$_2$ catalysts were prepared by a sol–gel method using tetraethyl orthosilicate (TEOS) and tantalum (V) ethoxide as the sources of silicon and tantalum, respectively. Two families of quaternary ammonium salts were used as surfactants, $[C_nH_{2n+1}(CH_3)_3N]$ Br ($n = 14, 16, 18$) and $[(C_nH_{2n+1})_4N]$ Br ($n = 10, 12, 16, 18$), with the catalysts that formed were denoted as Ta-n and Ta-4-n, respectively. Tantalum on MCM-41 (Ta-MCM) was prepared using an MCM-41 support. Deposition of tantalum was carried out by grafting from an alcoholic solution of tantalum (V) ethoxide. All catalysts contained 15 wt% Ta$_2$O$_5$ as determined by ICP-AES.

As expected, the binding energies corresponding to Ta $4f_{7/2}$ level exhibit typical values for Ta^{5+} for Ta-n, Ta-4-n, and Ta-MCM catalysts. A significant difference between the bulk and surface Ta/Si atomic ratios was found (refer Table 7.2) with materials showing surface enrichment of tantalum in all cases. In addition, for both Ta-n and Ta-4-n catalysts, as the chain of the directing agent was lengthened, the surface Ta/Si ratio also increased, with those formed from $[(C_nH_{2n+1})_4N]$ Br showing significantly higher ratios than those prepared using $[C_nH_{2n+1}(CH_3)_3N]$ Br. It is not clear whether the increase in the Ta/Si ratio observed is due to the decreased tantalum found in the pores of the catalyst or to the higher dispersion of tantalum over the surface, resulting in smaller Ta$_2$O$_5$ clusters. Both scenario would result in a higher Ta/Si ratio. The slightly higher Ta/Si XPS atomic ratio found for Ta-MCM cannot be compared with the ratios found for the Ta-n or Ta-4-n catalysts because of the significant difference in preparation method. For example, the grafting method used in the synthesis of Ta-MCM is more likely to lead to multilayer of tantalum oxide compared with the sol–gel method.

Table 7.2 XPS binding energies of Ta $4f_{7/2}$ and Si $2p_{3/2}$ levels and comparative XPS and chemical Ta/Si atomic ratios determined by ICP-AES (ref. 34)

Catalyst	XPS binding energy (eV)		Ta/Si atomic ratio × 10³	
	Ta $4f_{7/2}$	Si $2p_{3/2}$	Chemical analysis	XPS
Ta-14	26.4	103.2	1.15	6.4
Ta-16	26.4	103.2	1.15	6.9
Ta-18	26.5	103.2	1.15	7.5
Ta-4-10	26.4	103.2	1.15	10.2
Ta-4-12	26.5	103.2	1.15	13.4
Ta-4-16	26.4	103.2	1.15	18.8
Ta-4-18	26.4	103.2	1.15	19.4
Ta-MCM	26.5	103.4	1.15	23.4

7.4.1.6 *Effect of Promoter*

The performance of TiO_2 modified Pd catalysts containing TiO_2 either as an additive or support in acetylene hydrogenation is well studied. The origin of the selectivity promotion by the Ti species can be understood from the following XPS results. The peak representing the binding energy of Pd $3d_{5/2}$ appears almost at the same position for Pd/SiO$_2$/573 K and Pd – Ti /SiO$_2$ / 573 K but is shifted to lower binding energy for Pd – Ti /SiO$_2$ / 773 K. The peak shift is larger for Pd–10 Ti/ SiO$_2$ / 773 K. Charging effects have been minimized using an electron gun shower and the observed peak positions have been adjusted in reference to the C 1s peak. The shift in binding energy is influenced not only by the electronic interaction of the metal with other components but also by the size of the metal crystallites. Takasu *et al.* [36] observed in their study of Pd particles which were deposited on SiO$_2$ showed that the binding energy of Pd $3d_{5/2}$ decreased by 1.6 eV when the Pd particle size increased within a range smaller than 5 nm. The average Pd particle sizes of Pd/ SiO$_2$/573 K and Pd/ SiO$_2$/773 K are estimated to be 23 and 33 A° respectively, based on the H/ Pd ratio. In other words, the peak is shifted to lower binding energy because Pd particles grow after reduction at 773 K. However, the above interpretation does not explain the large peak shift observed for Pd–10Ti/SiO$_2$/773K because the shift should be to the same extent as that for Pd– Ti/ SiO$_2$ /773 K if it is solely dependent on Pd particle size. The XPS results also allow us to estimate the relative concentrations of Ti and Pd on the catalyst surface. The Ti/Pd ratio for the Pd–Ti/SiO$_2$ catalyst increases when the reduction temperature is raised from 573 to 773 K, suggesting that the Pd surface is covered with larger fractions of the Ti species after reduction at 773 K. The Ti/Pd ratio is significantly large for Pd – 10Ti / SiO$_2$ /773 K, which contains an excess amount of Ti species.

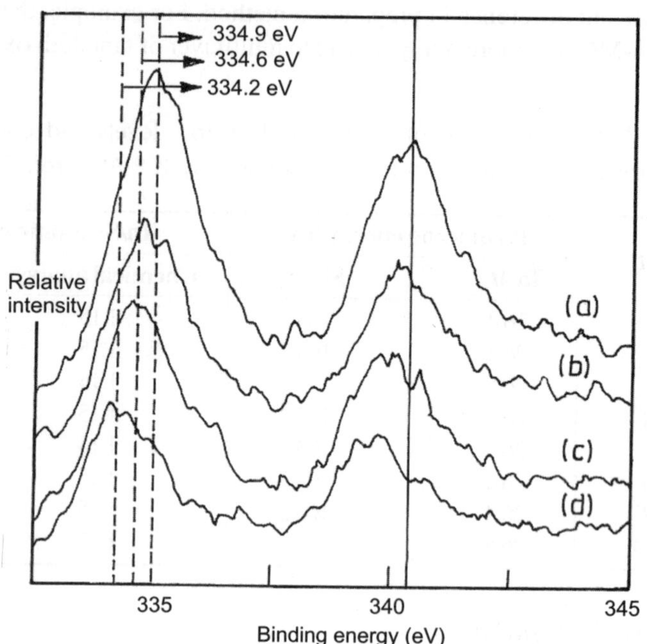

Fig. 7.30   XPS of Pd $3d_{5/2}$ for sample catalysts. (a) Pd/SiO$_2$/ 573 K, (b) Pd–Ti/SiO$_2$/573 K, (c) Pd–Ti/SiO$_2$/773 K, (d) Pd–10Ti/SiO$_2$/773 K (ref. 35)

7.4.1.7 Distribution of the Various Oxidation States of the Active Components

X-ray photo electron spectroscopy has also been used extensively for elucidating a number of properties of materials like polymers, oxidic materials and other inorganic solids. Both supported systems of active species as well as in oxidic materials, the oxidation states, density of states and cation distributions of the transition metals in various types of geometric interstitials can be explained by the XPS studies.

(a) Supported Catalyst Systems

Transition metal nitrides and carbides have been extensively studied because their catalytic properties resemble those of noble metals. The incorporation of nitrogen or carbon atoms into the lattice of a transition metal leads to an increase of d electron density at the Fermi level, and the property of the metal shifts towards those of noble metals. The similarity in chemical properties between transition metal nitrides and noble metals suggests that the former could be catalytically effective for NO elimination. A series of zeolite (H-ZSM-5)-supported molybdenum nitride catalysts with Mo loading ranging from 2 to 30 wt.% were synthesized and nitrided in a flow of ammonia. With increase of Mo loading from 2 to 30 wt. %, the degree of nitridation increased linearly with increase of Mo loading. It is observed that a catalyst with higher Mo loading exhibited higher initial activities. The nitrided 2 wt. % Mo/H-ZSM-5 catalyst was the most stable. NO conversion to N$_2$ remained unchanged within a test period of 15 h. For the catalysts with Mo loading above 2 wt.%, catalytic activity decreased with time on stream. After 15 h, the nitrided 2 wt.% Mo/H-ZSM-5 catalyst was the most active among the tested catalysts.

The Mo 3d spectra of the nitrided 2–30 wt.% Mo/H-ZSM-5 samples measured by XPS are shown in Fig. 7.31. The relative intensities of spin-orbit doublet peaks are given by the ratio of their respective degeneracy, and the I $(3d_{5/2})$ /I $(3d_{3/2})$ intensity ratio for the Mo $3d_{5/2}$–Mo $3d_{3/2}$ doublet is 3/2. A splitting of ~ 3.13 eV is expected for the doublet. By means of curve-fitting, the distribution of molybdenum oxidation states was estimated. There are two oxidation states of surface Mo, the peak at binding energy of 232.6 ± 0.2 eV is assigned to the Mo $3d_{5/2}$ component of MoO_3, while that at 229.1 ± 0.2 eV is assigned to the Mo $3d_{5/2}$ component of Mo_2N. The value (229.1 ± 0.2 eV) is higher than that of Mo^{2+} (228.5 ± 0.2 eV) but slightly lower than that of Mo^{4+} (229.6 ± 0.2 eV). Accordingly, the Mo species of Mo_2N is denoted as $Mo^{\delta+}$, where $2 < \delta < 4$. The XPS analysis of the Mo 3d binding energy shows that even at a Mo loading of 2 wt.%, there is nitrided molybdenum (Mo_2N) coexisting with MoO_3 on the surface of the zeolite. On the other hand, the proper deconvolution of the broad Mo 3d spectra of nitrided molybdenum samples, the XPS technique can provide information on the oxidation states of molybdenum species. Olivers et al. [38] studied the distribution of surface molybdenum ions as related to molybdenum loading. Based on the results of XPS measurements, and correlating Mo loading with the degree of nitridation, one can see a linear dependence between the two parameters. The Mo^+/ $(Mo^{2+} + Mo^{6+})$ ratio increased from 0.27 to 0.68 when the Mo loading increased from 2 to 30 wt. %. At a Mo loading of 2 wt.%, 27 % of the molybdenum was nitrided. With increase of Mo loading on the zeolite, there is a clear rise in the degree of nitridation. At a Mo loading of 30 wt.%, 65% of molybdenum was nitrided. One can deduce that the strong interaction between Mo species and the zeolite support hinder the complete nitridation of the supported molybdenum oxide. By curve-fitting the Mo 3d profiles of the nitrided 2–30 wt.% Mo/H-ZSM-5 catalysts, the degree of nitridation was estimated (as $Mo^{\delta+}/(Mo^{\delta+} + Mo^{6+})$ ratio); the results are summarized in Table 7.3. When the Mo loading on the zeolite was increased from 2 to 30 wt.%, the value of $Mo^{\delta+}/(Mo^{\delta+} + Mo^{6+})$ increased from 0.27 to 0.68, indicating that the degree of nitridation increased with increasing Mo loading.

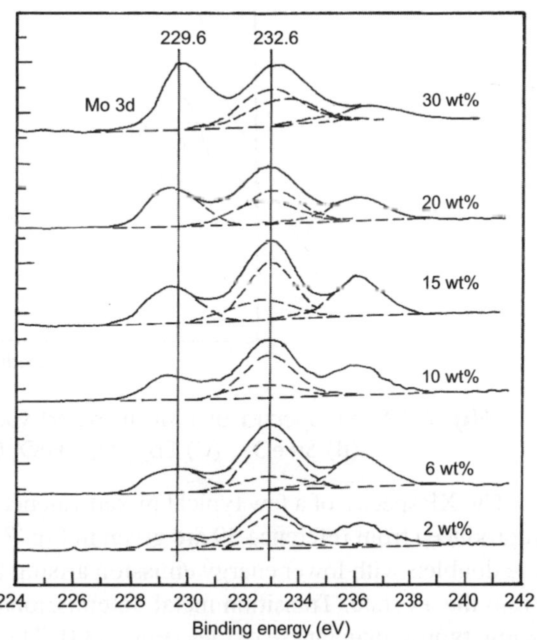

Fig. 7.31 Mo 3d spectra of the nitrided 2–30 wt.% Mo/H-ZSM-5 catalysts recorded in XPS study (ref: 37)

Table 7.3 Mo $3d_{5/2}$ binding energies and ratios of $Mo^{\delta+}/(Mo^{\delta+} + Mo^{6+})$ measured by XPS for the nitrided 2–30 wt.% Mo/H-ZSM-5 samples

Mo loading (wt.%)	Mo $3d_{5/2}$ Mo$_2$N (Moδ^+)	MoO$_3$ (eV)	$Mo^{\delta+}/(Mo^{\delta+} + Mo^{6+})$
2	229.2	232.6	0.27
6	229.2	232.6	0.32
10	229.0	232.5	0.37
15	229.1	232.6	0.42
20	229.1	232.6	0.53
30	229.2	232.7	0.68

(b) Transition Metal Ions in Oxides

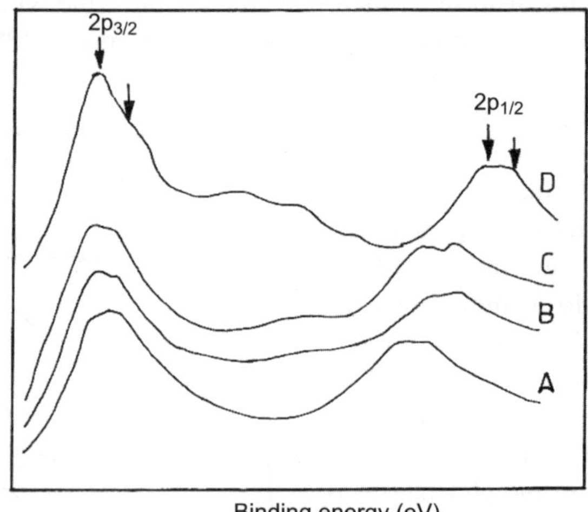

Fig. 7.32 XP spectra of typical mixed valence states of oxide ion (A) La$_{0.5}$Co$_{0.5}$ MnO$_3$ (B) SrFeO$_{3\text{-}8}$ (C) Eb$_{0.6}$ Sr$_{0.4}$ FeO$_3$ (D) Nd$_{0.5}$Sr$_{0.5}$ CoO$_3$ (ref. 39)

The XP spectra of a few typical mixed valence oxides (substituted LaMnO$_3$, SrFeO$_3$ and SrCaO$_3$) reproduced from reference 39 are given in Fig. 7.32. It is seen that both the 2p$_{1/2}$ and 2p$_{3/2}$ emissions are doublets with lower energy emission arising from the +3 state and the higher energy line arising from the +4 state. Transition metal 3p emissions also show +3 and +4 ions distinctly as shown for a few typical manganese oxides (Fig. 7.33). The valence bands also show the presence of multiple oxidation states although the assignments are not as straight forward due to closeness of bands arising from different final state effects. In Table 7.4, the values of core level binding energy of a number of mixed valence oxides along with those of some reference oxides are assembled to show how well XPS can identify the different oxidation states. This is particularly true for cobaltites since oxidation state Co^{3+} and Co^{4+} which could not be identified even by MossBauer spectroscopy, could be identified by XPS which has a time scale of 10^{-16} sec wherein complications due to electron hopping do not arise.

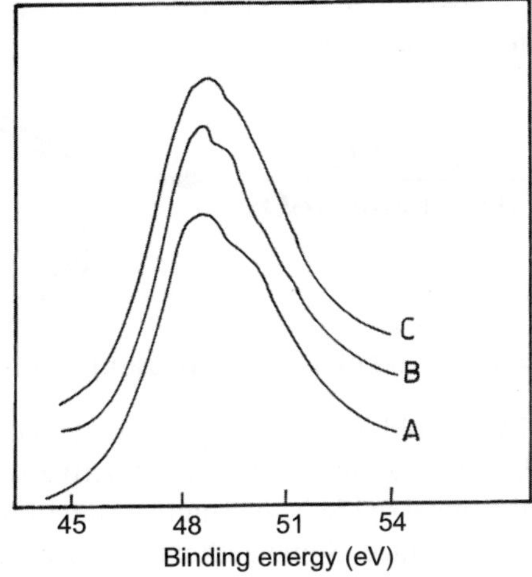

Fig. 7.33   Mn 3p bands showing 3+ and 4+ states (A) $LaMnO_3$ (B) $La_{0.5}Co_{0.3}$ MnO_3 (ref. 39)

Table 7.4   Core level binding energies (eV) of Mn, Fe and Co in mixed valence Oxides (ref. 29)

Compound		$2p_{3/2}$	$2p_{1/2}$	$3p$
$LaMnO_3$	Mn^{3+}	640.9	652.9	48.3
	Mn^{4+}	642.1	654.1	49.4
$La_{0.7}Ca_{0.3}MnO_3$	Mn^{3+}	640.8	652.7	48.2
	Mn^{4+}	641.8	653.8	49.4
$La_{0.5}Ca_{0.5}MnO_3$	Mn^{3+}	641.1	652.6	48.4
	Mn^{4+}	642.0	653.7	49.4
Mn_2O_3		641.2	652.9	48.3
MnO_3		641.9	653.5	49.5
$SrFeO_{3-x}$	Fe^{3+}	710.0	723.2	55.4
	Fe^{4+}	711.0	724.2	56.2
$Eu_{0.6}Sr_{0.4}FeO_4$	Fe^{3+}	709.6	723.1	54.9
	Fe^{4+}	710.8	724.0	56.0
Fe_2O_3	Fe^{3+}	710.3	723.6	55.2
$SrCoO$	Co^{3+}	779.8	795.3	60.4
	Co^{4+}	780.6	795.8	61.2
$Nd_{0.6}Sr_{0.4}CoO_3$	Co^{3+}	780.0	795.0	60.6
	Co^{4+}	781.0	796.0	61.6
$NdCoO_3$	Co^{3+}	780.1	795.5	60.5

7.4.1.8 *Electronic Structure of the Active Species*

Nanometer Au particles deposited on different supports of mesoporous Al_2O_3 via Homogeneous Deposition Precipitation method (HDP) has been used as highly active/selective and reusable catalysts for epoxidation of styrene. The dispersion and average size of the gold particles are dependent on the number of surface basic sites on the supports.

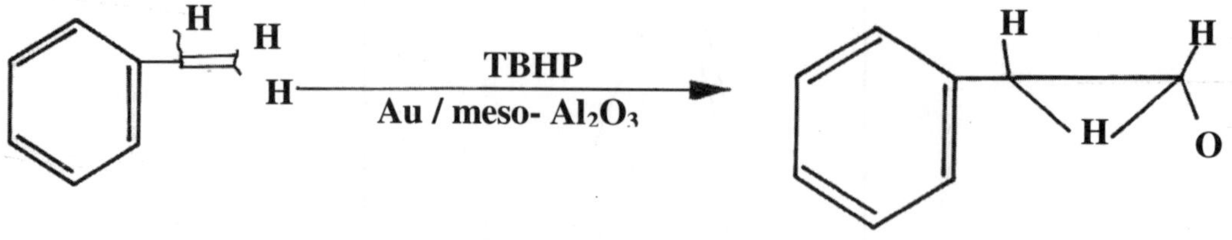

Scheme 1 Styrene epoxidation reaction using the Au/Al_2O_3 catalyst

Although the catalytic activity of gold catalysts in the low temperature CO oxidation has been intensively studied during the last decade, the nature of the active species is still discussed. It has been suggested that the role of the metal oxide is the stabilization of the gold nanoparticles and that the reaction takes place on the gold surface. Some authors proposed that the reaction takes place at the gold/metal oxide interface and that the metal oxide could act as a source of oxygen. Also the electronic structure of gold in active catalysts is unclear, but most of other authors suggest metallic gold to be the active species. The XPS spectra of Au 4f region (Au $4f_{7/2}$ and Au $4f_{5/2}$) are employed to investigate the state of gold on the different supports. The results indicated that only a binding energy of 83.8 eV for the Au $4f_{7/2}$ emission and 87.5 eV for the Au $4f_{5/2}$ electron ($\Delta E = 3.7$ eV) with the change of supports for the fresh samples. The binding energy indicates exclusively $Au^{(0)}$-particles without any trace of higher oxidized species on the four kinds of Al_2O_3 surface. The surface basicity of the support has no effect on the state of gold if the Au catalysts prepared under the conditions of HDP method [37].

7.4.1.9 *Synergetic Enhancement of Activity by Promoter and Support*

The nature of surface cesium compounds in cesium-modified ruthenium-sibunit (a new class of porous carbon-carbon composite materials combining advantages of graphite (chemical stability and electric conductivity) and active coals (high specific surface area and adsorption capacity)) catalysts for ammonia synthesis was studied by X-ray photoelectron spectroscopy. It was found that, on the reduction of promoted catalysts, cesium was incorporated into the micropores of sibunit to form quasi-intercalation cesium-carbon bonds. In this case, the chemical state of cesium was close to its state in cesium sub oxides. The subsequent interaction with atmospheric oxygen resulted in the oxidation of cesium, which occurred as cesium peroxide and cesium superoxide in the oxidized samples. Ruthenium occurred in a metallic state in the reduced samples. The activity of a $Ru-Cs^+/C$ (1) sample was higher than that of inactive $Ru-Cs^+/C$ (2). This is a consequence of the higher surface concentration of ruthenium, which is most likely due to an increase in the dispersity of metal particles, as well as of the higher probability of the interaction between the promoter and the active component due to a symbiotic increase in the surface concentrations of both ruthenium and cesium [40].

7.4.1.10 Metal Support Interaction

(a) Nickel-Alumina System

Catalysts containing about 8 wt% of the metal, prepared by impregnation of nickel nitrate onto, alumina were calcined at different temperatures (873 – 1673 K) in air for 6 h. Hydrogen absorption was rather difficult to measure on majority of the samples. XPS measurements were carried out on these samples and the binding energy value along with full width at half maximum are listed in Table 7.5. All the samples have Ni $2p_{3/2}$ primary peak around 857 eV as well as satellite peaks around 863 eV. Even after reduction there was no appreciable change in the intensity or binding energy values of these samples or FWHM. This indicated that there was no new nickel species formed after reduction. Comparison of the binding energy values of these samples with the standard values for Ni^0, NiO and $NiAl_2O_4$ suggests the presence of nickel aluminate, (surface spinel) which is difficult to reduce in all the calcined samples. Apart from the surface nickel aluminate, bulk spinel was also formed as was evidenced from XRD. This may suggest the interaction between nickel and alumina support when calcined at temperatures greater than 873 K.

Table 7.5 Binding energy values of Ni/Al_2O_3 calcined in air at different temperatures [41]

Sample	Temp K	Ni $2p_{3/2}$	Sat	DE_{sat}	O 1s	Al 2p
1	873	856.6 (3.8)	863.6	7.0	531.4 (3.6)	74.2 (3.0)
2	1073	857.4 (3.7)	864.2	6.8	532.2 (3.6)	75.2 (2.8)
3	1273	856.3 (3.5)	863.4	7.1	531.2 (3.5)	74.2 (2.3)
4	1373	856.0 (3.8)	863.1	7.1	530.9 (3.4)	74.0 (2.7)
5	1473	856.6 (3.2)	863.2	6.6	531.9 (3.1)	74.9 (2.6)
6	1673	856.2 (3.8)	862.4	6.2	531.3 (3.1)	74.2 (2.5)
Ni^0	––	852.6 (2.4)	––	––	––	––
NiO [42]	––	854.2 (3.6)	861.0	6.8	533.6 (3.4)	––
$NiAl_2O_4$ [43]	––	857.0 857.2	863.1 ––	6.1 ––	530.6 ––	–– ––

Samples 1-6 refer to calcinations at different temperatures given in column 2. Values in the parenthesis represent FWHM.

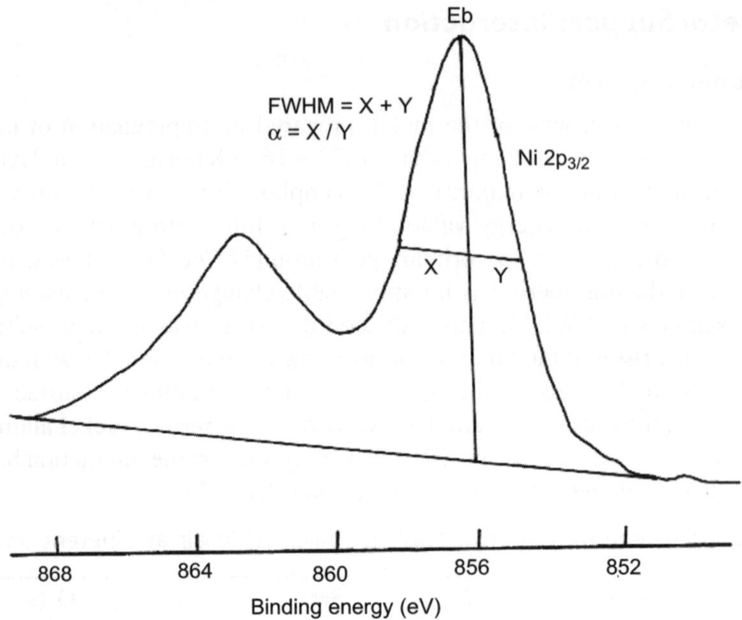

Fig. 7.34 Ni 2p parameters used in this study, E_b binding energy; FWHM = peak width at half maximum, a = asymmetry factor (ref. 29)

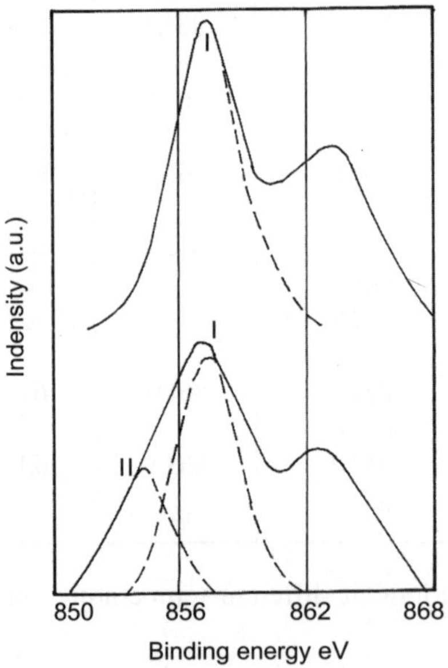

Fig. 7.35 XP spectra of a fresh and reduced nickel mordenite (ref. 43)

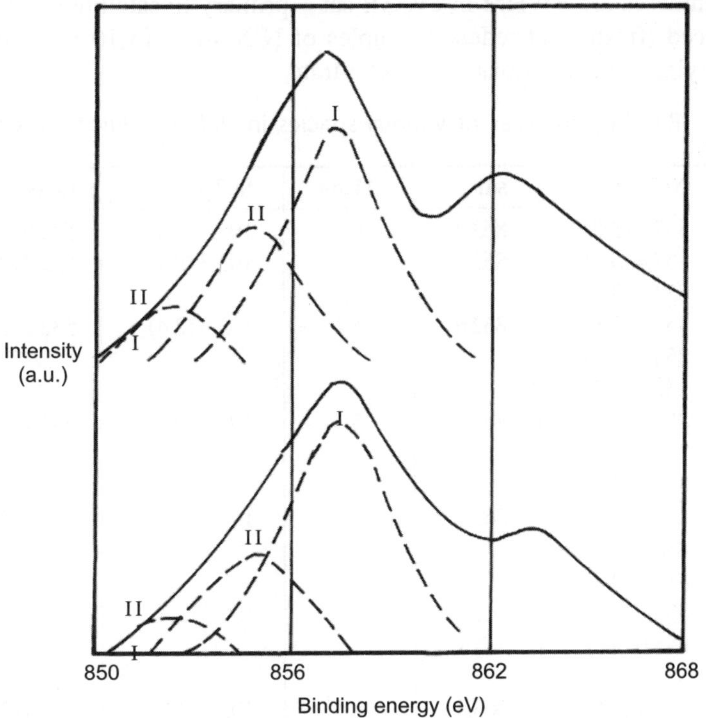

Fig. 7.36 XP spectra of reduced samples of NiNaM (ref. 43)

(b) Nickel Mordenite

Ni^{2+} ion in zeolite system is not easily reducible and it depends on several factors. It is rather difficult to get a homogeneous dispersion of the metal. The problem of reduction of Ni ions incorporated by ion-exchange in mordenite (zeolite) can be understood with the help of XPS studies. The XPS spectra of fresh (unreduced) and reduced nickel mordenite are shown in Fig. 7.35. The fresh catalyst has a primary peak at 857.2 eV with FWHM of 3.8 eV and a satellite peak at 863.2 eV. On reduction, the primary peak broadened with the appearance of a shoulder around 854 eV, indicating a new species.

Deconvolution of the peak shows two species of Ni, peaks I and II, corresponding to binding energy 857.4 eV (FWHM, 3.8 eV) and 854.2 eV (FWHM, 3.4 eV) respectively. A perusal of the XP spectrum in Fig. 7.36, for other reduced samples of NiNaM also shows that spectra are broad and unsymmetrical with FWHM greater than 6 eV. Deconvolution of the peaks and comparison of the binding energies for various nickel species revealed that a multiple species of Ni (Ni^0, Ni^+, Ni^{2+}) are present in the reduced sample. It is also clear that the contribution of the Ni species to the total spectra is relatively small. XPS peaks along with hydrogen adsorption measurements on Ni mordenite catalysts confirm the difficulty of reduction of Ni^{2+} ions to Ni^0. When Ni^{2+} is supported on zeolite, incomplete reduction is responsible for low dispersion of the metal. The data used for the identification of various species on nickel mordenite by XPS after various reduction treatments are summarized in Table 7.6.

Values in parentheses are FWHM in eV. ΔE sat = primary to satellite line separation. F and R represent unreduced (fresh) and reduced samples of NiNaM. 3 to 10 refer to nickel mordenite prepared with varying concentrations of nickel nitrate.

Table 7.6 Binding energies of various species in different Ni- Mordenite catalysts

Catalyst	Ni $2p_{3/2}$	sat	ΔEsat	Si 2p	O 1s	Ref.
3 (F)	857.2 (3.8)	863.2	6.0	103.7 (2.4)	532.5 (2.4)	43
3 (R)	857.4 (3.8)	862.8	5.4	103.3 (2.4)	532.3 (2.4)	43
	854.2 (3.4)					
4(R)	857.2 (3.8)	862.6	5.4	103.3 (2.4)	532.2 (2.4)	
	854.8 (3.2)					43
	852.4 (2.2)					
5 (R)	857.4 (3.8)	863.2	5.8	103.4 (2.4)	533.1 (2.5)	
	855.0 93.2)					43
	852.6 (2.2)					
6 (R)	857.2 (3.8)	862.8	5.6	103.2 (2.4)	532.3 (2.4)	
	854.8 (3.2)					43
	852.4 (2.2)					
7 (R)	857.4 (3.8)	862.8	5.4	103.3 (2.4)	532.4 (2.4)	
	855.0 (3.2)					43
	852.2 (2.2)					
10 (R)	857.4 (3.6)	863.4	6.0	103.3 (2.4)	532.4 (2.4)	
	854.4 (3.2)					43
Ni (0)	852.8 (2.0)	858.6				44
NiO	854.2	861.0				
	854.8 (4.2)					43
	854.2				529.7	45
Ni_2O_3	856.9	862.8				46
$NiNl_2O_4$	855.2				530.6	45
	857.0	863.1				46
	857.2					47
Ni(OH)$_2$	856.0	862.2				43
	856.3 (4.1)					48
	855.6					

7.4.1.11 Surface Growth of the Active Metal

From successive XPS measurements it is possible to observe the over layer growth on the support as can be seen directly from the XPS spectrum given in Fig. 7.37. It shows typical variations of Pd $3d_{5/2}$ signal during the layer growth on the substrate. Beside the intensity enhancement, two effects can be seen. The binding energy E_B is shifted to higher values for smaller particles, and the FWHM of peaks decreases during the growth. FWHM parameter variations are usually associated with mean coordination number of cluster atoms.

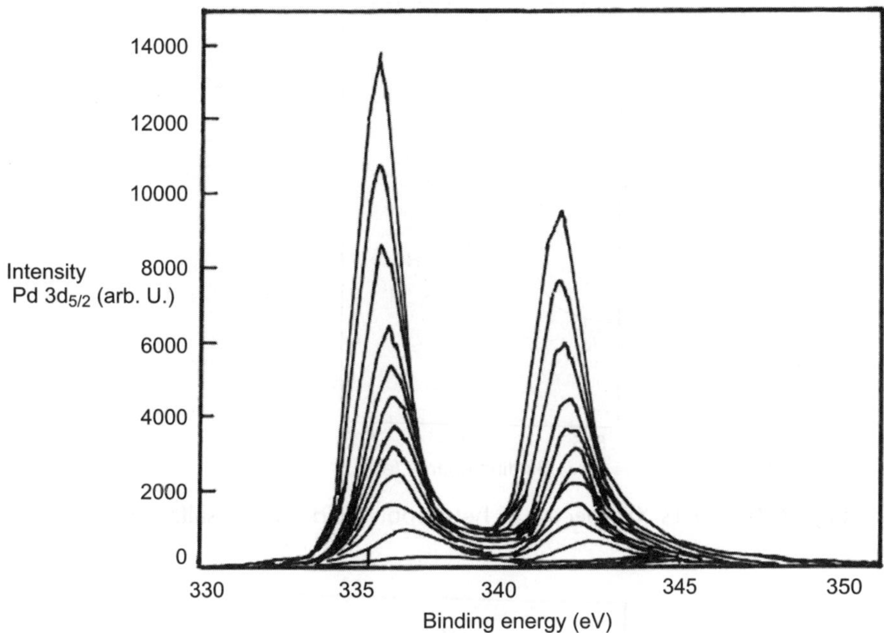

Fig. 7.37 Variation of the Pd $3d_{5/2}$ photoelectron signal during the Pd layer growth on ?-Al_2O_3 substrate (ref. 49)

7.4.1.12 *Acidity of the Catalysts*

12- Tungsto phosphoric acid and its salts were prepared and calcined at 573 K and their acid strength distributions were evaluated using visible Hammett indicators. In view of the observed gradation in the acidity of HPW and its salts, the oxygen 1s lines of these compounds was scanned and are shown in Fig. 7.38. The spectra obtained from the acid and its salts of Al and Na are given. It is seen that the oxygen 1s peak of the Na salt is at a lower binding energy as compared to that of the free acid or the aluminium salt. It is also to be noted that the oxygen 1s peak of sodium salt is asymmetric which could be due to the contribution from low binding energy oxygens which are more negatively charged as compared to the majority of other oxide ions of the $PW_{12}O_{40}^{3-}$ anion. Eventhough one could guess the least acidic nature of the Na salt from the absolute binding energy values of the oxygen 1s peak, it is not clear whether this approach can be used for comparative evaluation of acidity of a number of heteroploy acids as the observed shift and the asymmetry of the oxygen 1s peak is small. Similarly XPS peaks of 4f level of W are shown in Fig. 7.39. It can be concluded that though in principle the relative acidity of the salts of heteropoly acids should be reflected in the binding energy shifts of the cation of the salt and oxide ions, these shifts are too small to be monitored directly. The relative acid strength distribution between the various salts can probably be deduced from XPS measurements only when one considers the difference between the peak positions of the cation and anion. The difference between the oxygen 1s line and the tungsten 4 f line was found to be 495.5 eV for the free acid, 495.4 eV for the aluminium salt and 495.1 eV for the sodium salt which reflects the order of acidity of these substances.

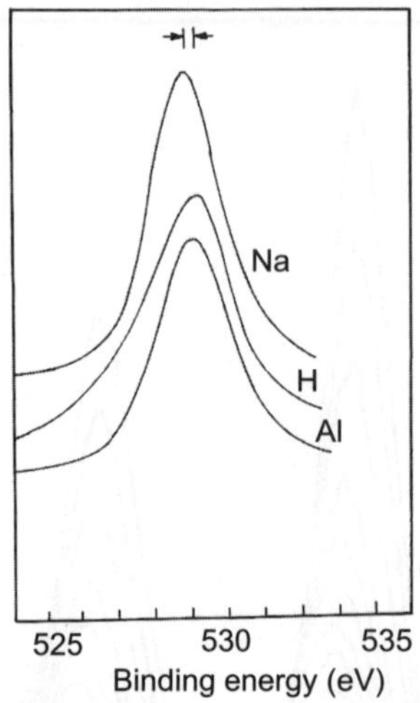

Fig. 7.38   O 1s XP spectra of heteropoly acid and its salts (ref. 50)

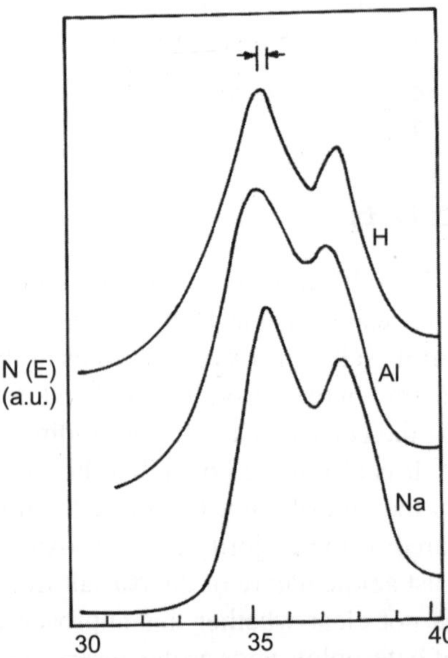

Fig. 7.39   W-4f XP spectrum of heteropoly acid along with spectra of its Al and Na salts (ref. 50)

7.4.1.13 *Cation Distribution in Spinels*

Nickel magnetite annealed at 673 K in vacuum can have either of the two cation distributions.

(1) $(Ni^{2+}_{0.1} Mn^{3+}_{0.9})_t [Ni^{2+}_{0.9} Mn^{3+}_{1.1}]O_4$

(2) $(Ni^{2+}_{0.1} Mn^{2+}_{0.9})_t [Ni^{2+}_{0.9} Mn^{3+}_{0.2} Mn^{4+}_{0.9}]O_4$

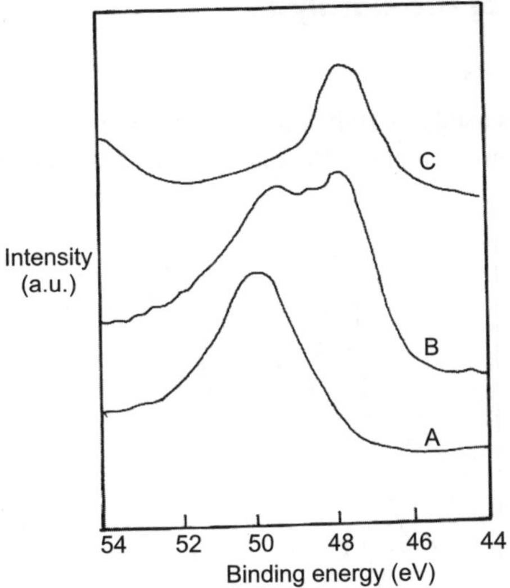

Fig. 7.40 Comparison of Mn 3p of $NiMn_2O_4$ (B) with that of $MnFe_2O_4$ (C) and $NiMnZnO_4$ (A) (ref. 51)

In configuration (1); trivalent manganese is present in tetrahedral as well as octahedral sites and the only effect on the core level may be caused by the site differences. The effect of site difference upon binding energies of core levels is small and may be of the order of a few tenth of an electron volt. In configuration (2), the main portion of the octahedral manganese ions is tetravalent and the ions at the tetrahedral sites are divalent which may result in the shift of binding energies by about 2 eV. As the ligands are oxygen ions for both the lattice sites, and since the covalency effects are supposed to have only minor effect, the core levels of Mn^{2+} and Mn^{4+} ions must be distinguishable. The 3p levels of manganese in $NiMn_2O_4$ are shown in Fig. 7.40 together with those of $MnFe_2O_4$ and $NiMnZnO_4$. The separation into $3p_{1/2}$ and $3p_{3/2}$ is not seen in the ferrite or in the zinc compound, although the energy difference is of the order of 1 eV. The Mn 3p lines have a FWHM of 1.8 and 2.8 eV for the ferrite and zinc compound respectively. In the case of zinc compound, the larger FWHM value could be caused by the reduction of some manganese ions in the spectrometer. However the majority of the Mn ions in $MnFe_2O_4$ and $MnNiZnO_4$ are supposed to be di and tetravalent respectively. Comparing the Mn-3p spectra in $NiMn_2O_4$ with those in $MnFe_2O_4$ and $MnNiZnO_4$, one can say that the $NiMn_2O_4$ spectrum is a superposition of the other two spectra. This shows the simultaneous presence of Mn^{2+} ions (tetrahedral as in $MnFe_2O_4$) and octahedral Mn^{4+} ions (as in $NiMnZnO_4$). The binding energy values for the 2p levels for Mn^{4+} ions in $NiMn_2O_4$ and $NiMnZnO_4$ were shifted by 1.5 to 1.9 eV to higher values with respect to the 2p levels of the Mn^{2+} ions in other spinels. The $2p_{3/2}$-$2p_{1/2}$ separation has a value 12 eV, though splitting of lines due to the presence of Mn^{2+} and Mn^{4+} was not observed. The presence of these ions (Mn^{2+} and Mn^{4+}) in $NiMn_2O_4$ could be deduced from the FWHM values (4.5 eV) as compared to a value of 3 eV for other compounds where there is no combination of Mn^{4+} and Mn^{2+} ions.

7.4.1.14 *Deactivation of the Catalysts*

(a) Oxychlorination Catalysts

Copper chloride on alumina is used as catalyst for oxychlorination of ethylene. The oxidation states of copper in the catalyst system influences the catalytic activity. The reason for catalyst activation

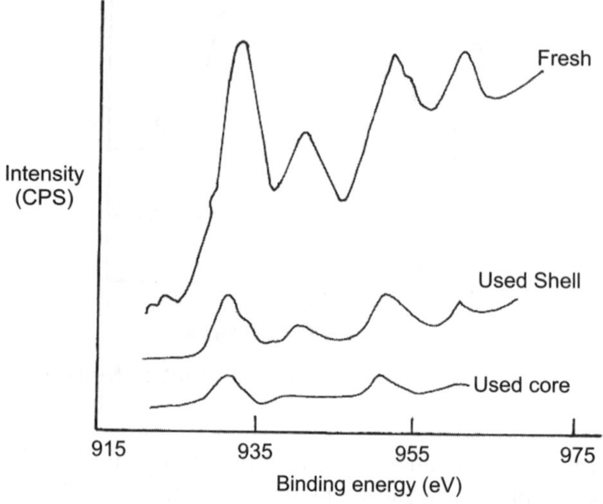

Fig. 7.41 XP spectra of oxychlorination catalyst in region of copper (ref. 52)

was monitored by analyzing and comparing binding energy values of copper in fresh and spent catalysts. Cu^{2+} species show typical satellite due to shake up electrons associated with $2p_{1/2}$ and $2p_{3/2}$ energy levels, whereas these satellites are absent for Cu^+ species. The absence of satellite indicates the absence of Cu^{2+} species, whereas the presence of them indicates copper to be in a mixture of +2 and +1 states or +2 states only. A look at the XP spectra of the catalysts in the region of Cu $2p_{1/2}$ and Cu $2p_{3/2}$ binding energies and comparison with that of the pure compounds suggests that the core of the spent catalyst contains only Cu^+ phase as indicated by the absence of satellite , whereas the shell contains Cu^+ and Cu^{2+} species. The copper binding energies of 932.8 eV ($2p_{3/2}$) and 952.8 eV ($2p_{1/2}$) correspond to Cu^+ from the compound $CuAlO_2$. Formation of such an irreversible phase under reaction conditions may upset the redox equilibrium between Cu^+ and Cu^{2+} states and contribute to the catalyst deactivation.

(b) ZSM-5-MTG Catalysts

ZSM-5 is an important catalyst for the conversion of methanol to gasoline and carbon deposits gradually during the reaction and deactivates the catalyst to the point where it becomes inactive. ZSM-5 has an internal channel with a surface area of about 400 m^2/g and an external area of 5 m^2/g. This catalyst has been deliberately coked with carbonaceous residues to various loadings in the range 0-17 wt% of C by reacting with methanol feed for different durations. If carbon deposits only internally then fractions of the carbon in the outer 5-10 nm of the crystallite would be visible by XPS. On the other hand, if C deposits on the external 5 m^2/g surface, XPS would detect the rapid build up of surface coke. As seen from Fig. 7.42 it is clear that below 10 wt% coke, the majority of zeolite crystals fill their internal pores with coke and above 15 wt%, carbon begins to precipitate on the external surface, resulting in a dramatic increase in surface carbon levels.

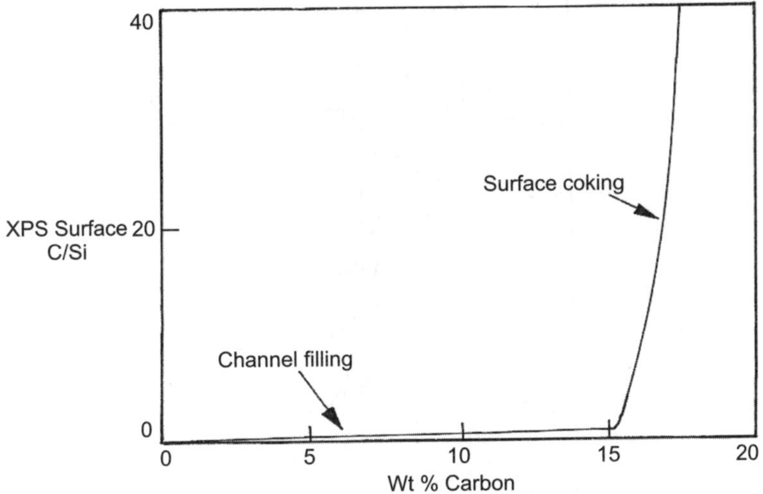

Fig. 7.42 XPS – derived surface C/Si ratios vs. coke loading for ZSM-5 zeolites, coked by reacting in a methanol feed for different durations. Comparison of calculated C/Si ratios for two different mechanisms: (a) Channel filling and (b) coking on the external crystal surfaces indicates that the channels become full at 15.0 ± 0.5 wt % C. Beyond this point surface coking predominates and the silicon signal is rapidly attenuated by the external carbon over layers (ref. 53)

7.4.1.15 Effect of Pretreatment of Catalysts

The effect of successive reduction treatments on a hydrodesulphurization catalyst, cobalt-molybdenum – alumina is shown in Fig. 7.43. The change in the Mo 3d spectra is very well pronounced as the time of the heat treatment increases. The air fired catalyst in spectrum 'a' contains entirely Mo (VI) at the surface, but with repeated hydrogen reduction the spectrum changes radically and can be curve resolved into Mo (VI) (V) and (IV) contributions. The chemical state of the surface in spectrum (f) contains approximately equal proportions of the three oxidation states [54].

7.4.1.16 Surface Oxidation

The oxidation of metallic surfaces by oxygen is illustrated by considering nickel as example. At low oxygen exposures, emission from a level characteristic of adsorbed oxygen is observed at 5 eV below the Fermi energy, in addition to the emission from the metallic d bands of the Ni within 3 eV of the Fermi energy. Further exposure results in the spectrum losing its metallic character and becoming insulator-like; in other words the density of states at the Fermi energy has nearly vanished. By comparison with the spectrum of bulk NiO in Fig. 7.44, one can identify the levels for oxidized Ni near 2 eV and 5 eV with contributions of the d-electrons of oxidized nickel and with surface oxide p bands. Thus the spectra show that oxidized Ni is similar to bulk NiO.

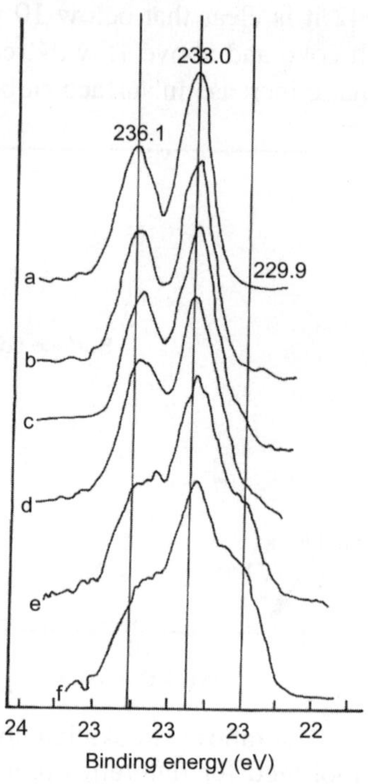

Fig. 7.43   The changes in Mo 3d XPS spectrum from a cobalt – molybdenum – alumina catalysts during successive reduction treatments in hydrogen at 500°C. (a) Air- fired catalyst; (b) Reduction time 15 min.; (c) 50 min.; (d) 60 min.; (e) 120 min.; (f) 200 min. (ref. 27)

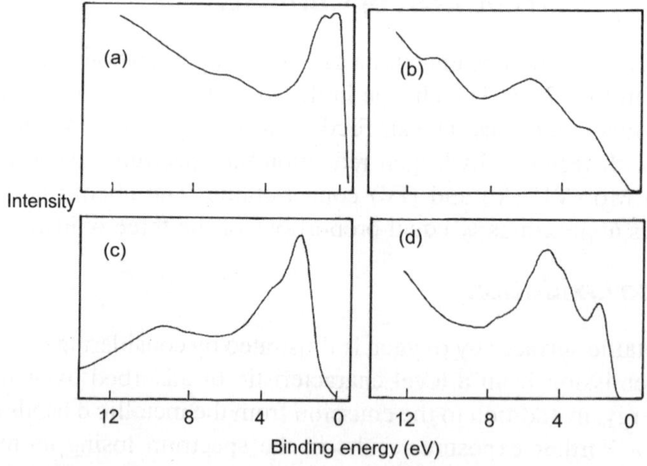

Fig. 7.44   (a) UPS spectra for clean Ni and for Ni with a half monolayer of chemisorbed oxygen (b) Spectrum of surface oxide formed by oxygen exposure. This spectrum shows features similar to that of bulk (100) – single crystal NiO (c) Ni^{2+} 3d state emission and oxide p-band emission (d) XPS for bulk NiO showing the Ni^{2+} 3d state emission (ref. 55)

7.4.1.17 *Study of Adsorption Phenomena*

The X-ray photoelectron spectra obtained for the adsorbed CO on Ni (100) at 77 K are shown in the Fig. 7.45 (a)Spectrum in this figure corresponds to the 2σ(C (1s)) level region and b) to 1σO ((1s)) level region. In both cases, in addition to the main emission at 286 eV and 532 eV respectively, shoulders appear at ~ 291 eV and ~ 537 eV with respect to Fermi level. The separation between the shoulder and main peak is almost 6 eV in both the cases. The corresponding spectra obtained for CO adsorbed on Ni (100) at 100 K are shown in the same Fig. 7.45 for two coverages namely 0, 3L, and 7.0 L of CO. The positions of the peaks are at 285.4 eV and 531.1 eV at 0, 3 L exposure, while they are at 285.5 eV and 531.2 eV at 7.0 L exposure. The shift in the binding energy with coverage is very small.

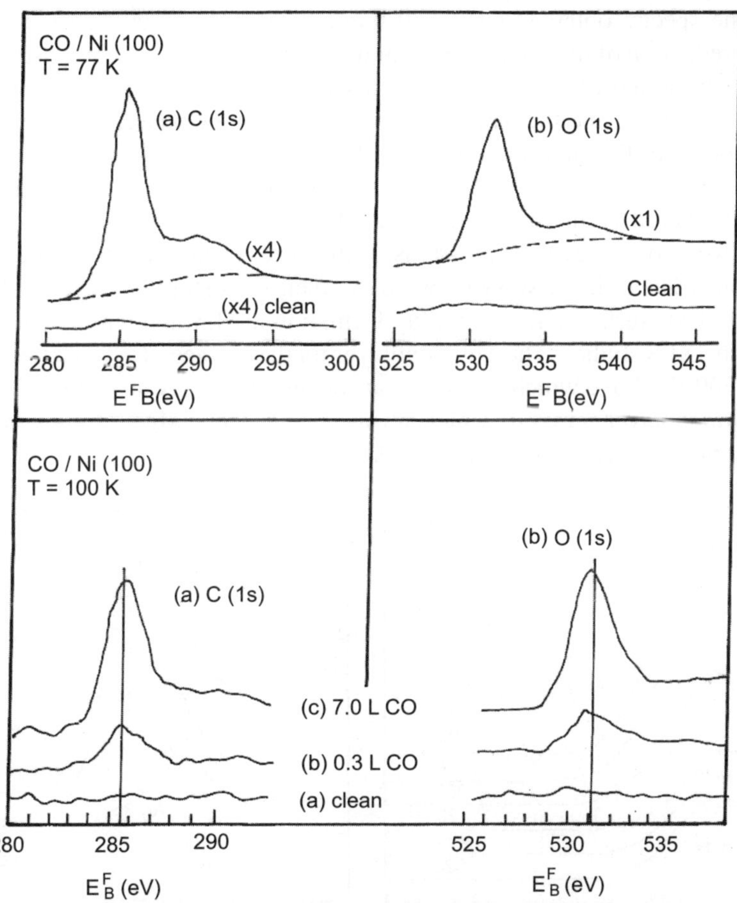

Fig. 7.45 X-ray photo electron spectra of adsorbed CO on Ni (100): (a) after adsorption at 77 K (ref: 56) b) after adsorption at 100 K (ref. 57)

The positions of the peaks due to 2σ or 1σ emissions are around ~ 286 eV and ~ 531 eV referenced to E_F. These positions are approximately 2 eV higher than those observed for atomic adsorption of C and O on these surfaces. In addition to these main peaks, peaks appear around 290 eV, which is considered to be due to shake – up process. The satellite structure observed in the core level excitations for adsorbed CO is usually considered in terms of screened and unscreened states. These situations are based on the assumption that the 2π level of CO is pulled down below the Fermi level of the adsorbent. Another possibility is that the satellite peak at ~ 6 eV with respect to the main core level emission peak is due to simultaneous excitations of the 5σ or 1π to 2π levels occurring together with the ejection of the photoelectrons.

7.4.1.18 Temperature Effects in Solid State Materials

It is possible that the electron generated by the source and heat produced can sometimes decompose the complex i.e., the organometallic complexes used as the precursors for catalysts. In such cases one has to interpret the spectra obtained taking all these secondary effects that can occur inside the spectrometer like reduction of the sample in vacuum, the charging, the decomposition of the species due to heat and electrons and all other side effects due to operating conditions. In supported catalysts depending on the dispersion and the particle size of the supported phase, the intensity ratio of I_p / I_s due to the emission from the supported phase and the support, can be small when the support is exposed and the dispersion is small or can be large if the dispersion is good and the support phase is not exposed and covered by the supported phase. This simple explanation will not hold good if the supported phases were to occupy and spread inside the pores of the support or the particles were to take different geometries, or if the supports were to function as stratified layers. Monolayer of Mg formed on the Ag (100) surface shows an alloy formation at elevated temperatures, which can be understood from the XPS studies. The Mg monolayer on the surface of Ag (100) is stable only upto a temperature of 300 K, above which a desorption sets in and finally at 420 K a Mg concentration equivalent to 0.5 monolayer remains, which forms a disordered Mg–Ag alloy with the topmost silver layers, as concluded from X-ray induced AES.

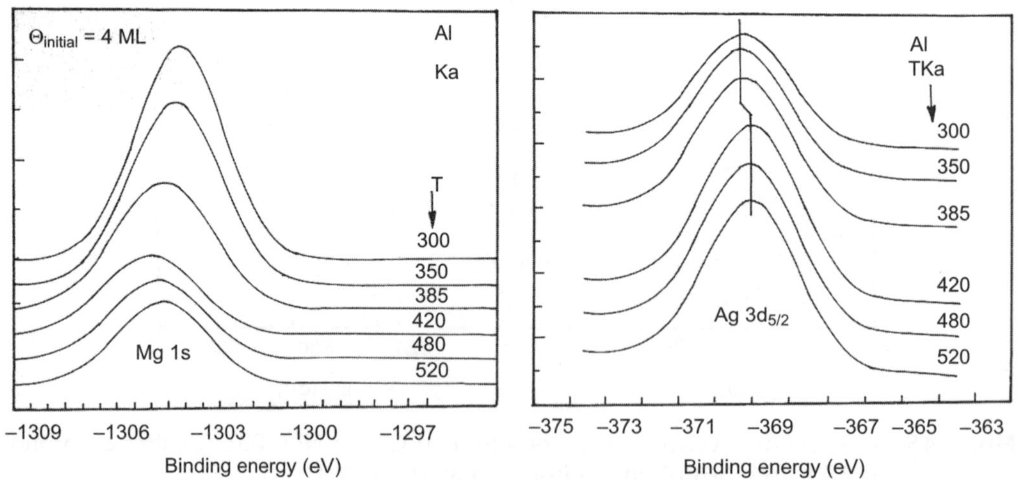

Fig. 7.46 Temperature dependent chemical shifts for Mg 1s and Ag 3d levels

Vanadium dioxide is an insulator at room temperature but undergoes a structural change at 339 K and becomes a metallic conductor. Fig. 7.47 shows that the gap created at the lower temperature, which results in the lower conductivity.

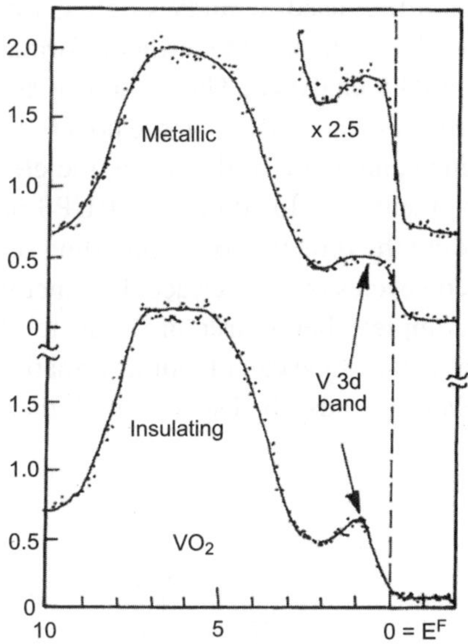

Fig. 7.47 XPS valence bands of VO$_2$ at temperatures above and below the metal-insulator transition temperature (ref. 24)

The microscopic origin and the mechanism stabilizing the low-temperature structural order is still a matter of debate. At metal insulator transition temperature (T_{MI} = 339 K) the structure changes on cooling from a metallic tetragonal rutile structure to an insulating distorted rutile structure with monoclinic symmetry. It is visible in the E_F region of the XPS, the change in density of states of VO$_2$ when it is undergoing a metal to insulator transition.

7.5 ULTRAVIOLET PHOTOELECTRON SPECTRA OF ATOMS

7.5.1 Introduction and Examples

The commonly used sources such as He I (21.2 eV) and He II (40.8 eV) are at low energies, only valence electrons can be ejected and this spectroscopy is restricted to probing bonding in metals, molecules and in adsorbed states. It is these electrons which form a chemisorption bond and knowledge of electronic density of states at a surface are of vital importance in attempts to understand the formation of chemical bonds between solid surfaces and adsorbed atoms or molecules. It is clear that the electron coming from the Fermi level has a kinetic energy E_k = hv $-$ ϕ and the slowest electrons with E_k = 0 with the highest binding energy since the kinetic energies of the electrons in UPS is low between 5 to 15 eV and since the final state of the photoelectron is still in the region of the unoccupied part of the density of states of the metal, this spectrum can be considered at joint density of states of both occupied and unoccupied states. If the photon energy were to increase then the final states will be in the range corresponding to entirely free electrons in vacuum. If the photon energy increases then one can probe the true density of sates of the metals. In XPS also one probes the true density of states of the metals, but it is at a lower resolution due to broader line width of the X-ray source. UPS is most often used to study single crystals and to probe the molecular orbitals of chemisorbed molecules. UPS can provide further information on the system if the emitted electrons are both energy and spatially analyzed. This is known as angle resolved UPS (ARUPS). Using ARUPS, the band structures of clean and adsorbate covered surfaces have been determined. By changing the angle of incidence and angle of detection, the electronic orbitals from which the photoelectrons are ejected can be identified. In addition, ARUPS provides detailed information on the surface chemical bond including the direction of the bonding orbitals and the orientation of the molecular orbitals of the adsorbed species on the surface. In general UPS can be used to obtain the electronic structure of solids (complete band structure) and to study the adsorption of simple molecules such as CO and N_2 on metals. One can obtain information on chemisorption bonding by compairing the molecular orbitals of the adsorbed species with those of both the isolated molecule and with model calculations [24].

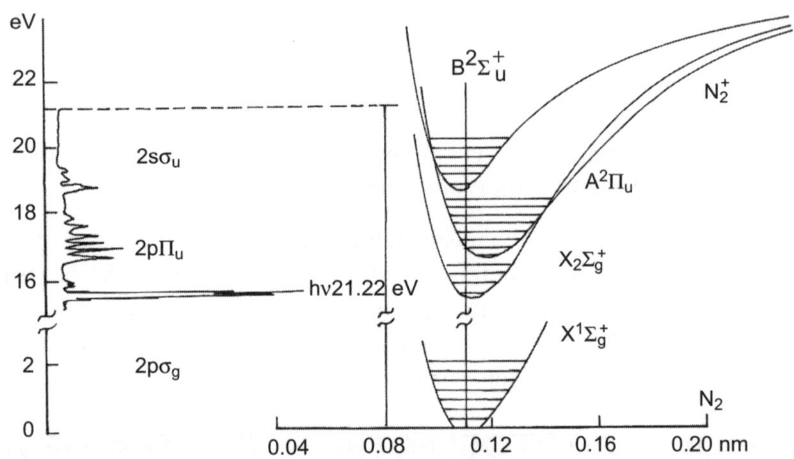

Fig. 7.48 Photo electron spectrum of N_2 (ref. 58)

Photoelectron spectrum of nitrogen molecule showing that it is a reflection of the potential energy curves.

The spectrum corresponds to three different transitions. These three different transitions correspond to three different electronic states of ionized nitrogen. The potential energy curves of these ionic states are shown in Fig. 7.48 with vibrational levels in them. Each of the peaks shows vibrational excitations. In fact the photoelectron spectrum is a reflection of the potential energy curves of the system. By 21.22 eV photon, it is possible only to access these three levels. If one has sufficiently energetic photons, it could have been possible to look at other deeper layers. From the photoelectron spectrum, it is also possible to get an idea on the kind of energy levels involved in the ionization process. These levels actually correspond to orbitals, which participate in bonding. Ionization from these levels can give information on the kind of bonding possible in the system. In this particular case, the first excitation corresponds to the removal of electron from a non-bonding orbital. The second corresponds to removal of an electron from a bonding orbital. The third corresponds to removal of an electron from a weakly anti-bonding orbital. It is clear that a vibrational spacing of the first ionic state is similar to that of the ground state. For the second one, the vibrational spacing is smaller than that of the ground state. For the third one, the vibrational spacing is larger than that of the ground state. If an electron is removed from a non-bonding orbital, obviously the bonding characteristics of the molecule are not disturbed. Therefore, inter-nuclear distance remains unchanged and vibrational separation is largely the same. If one removes an electron from a bonding orbital, the state that is created has less bonding character, inter-nuclear distance can increase and therefore, the vibrational quantum may be reduced. If one removes an electron from an anti-bonding orbital, the state that is created has more bonding character and therefore the vibrational quantum is larger. It is also reflected in the equilibrium internuclear distance which is unaffected in the first state and is shifted to the right in the second case and is shifted to the left in the third case. Thus photoelectron spectrum gives a qualitative idea of the kind of bonding.

7.5.2 Effect of He I and He II Radiation

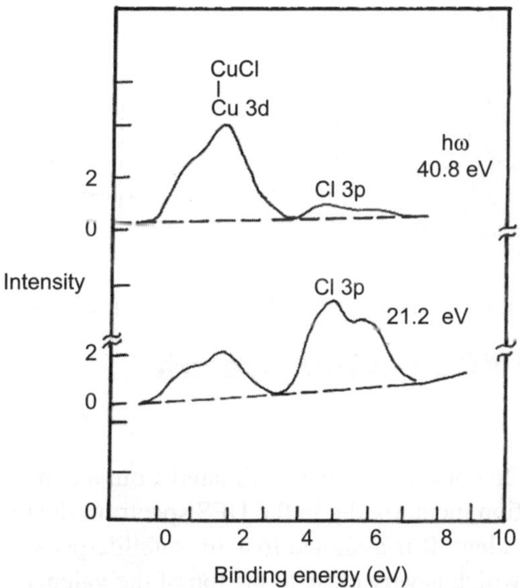

Fig. 7.49 Effect of He I and II on the UP spectra of CuCl (ref. 58)

In the photo excitation process using photons of two different frequencies with increase in final state energy an increase of overlap of the final state wave function with those of higher angular momentum initial states are expected (i.e., going from He I to He II radiation as shown in the Fig. 7.49 the overlap of valence bands increases). This is a useful property which can be employed to analyze complicated spectra. Fig. 7.49 presents a comparison spectra taken with He I (21.2 eV) and He II (40.8 eV) radiation for the valence band of CuCl. One sees that the signal from the Cu 3d emission increases relative to the signal strength from the Cl 3p electrons in going from He I to He II radiation.

7.5.3 Effect of Time on UP Spectra

It is still in doubt, whether this technique is non destructive or not. The spectrum of Cu (100) measured in equal time intervals between 0 and 10 hours after surface preparation with constant illumination of the sample during the time of the experiment, the signal increases in width and falls in intensity with time. This shows that the technique is in principle destructive. But at higher temperatures, there is no such deterioration of the signal with time; hence it is not clear whether it is due to radiation damage or due to the adsorption of residual gases. The corresponding spectra are given in Fig. 7.50.

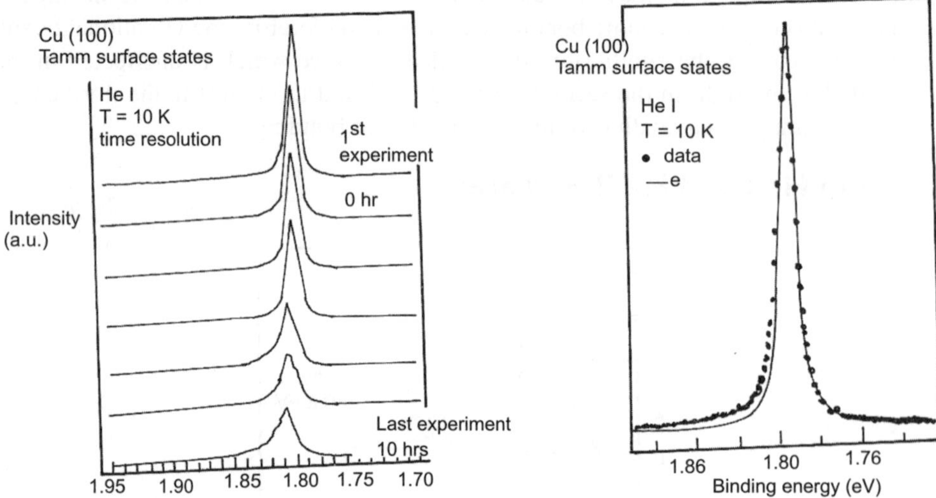

Fig. 7.50 Variation of UP spectra with time for Cu (100) at 10 K (ref. 58)

7.5.4 Comparison of XPS and UPS Spectra

(a) NiO

In the XPS spectrum, the O 2p contribution was estimated from a comparison of XPS spectra with that of the UPS of the transition metal oxide. In the UPS spectrum the O 2p band is visible. In both cases a distinct satellite is evident. It is assigned to a $3d^7$ satellite peak (final state). The zero is the experimental Fermi energy, which is very close to the top of the valence band. So by comparison of UPS and XPS it is possible to estimate quantitatively the passive oxide layers formed on the surface. The relevant data are given in the Fig. 7.51.

(b) CuO and Cu_2O

A comparative study of valence bands of polycrystalline samples of Cu_2O and CuO taken with He I, He II, Mg-K_α shows that in Cu_2O, O 2p and Cu 3d states are separated and with Mg-K_α in the XPS only metal 3d density of states are predominant. With UV sources, the spectrum gives a fair representation of the combined O 2p and Cu 3d density of states. In the case of CuO valence band studies in a similar way shows that the separation between the O 2p and Cu 3d orbitals is no longer so pronounced, indicating a stronger hybridization between the ligands and the metal d electrons. The data are given in the Fig. 7.52.

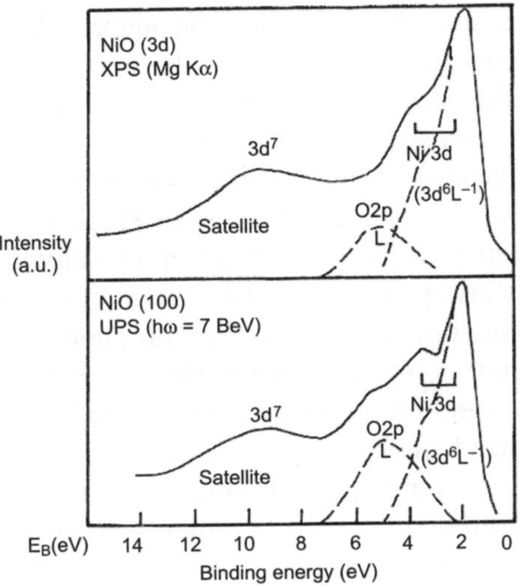

Fig. 7.51   Comparison of XPS and UPS spectra of NiO (ref. 58)

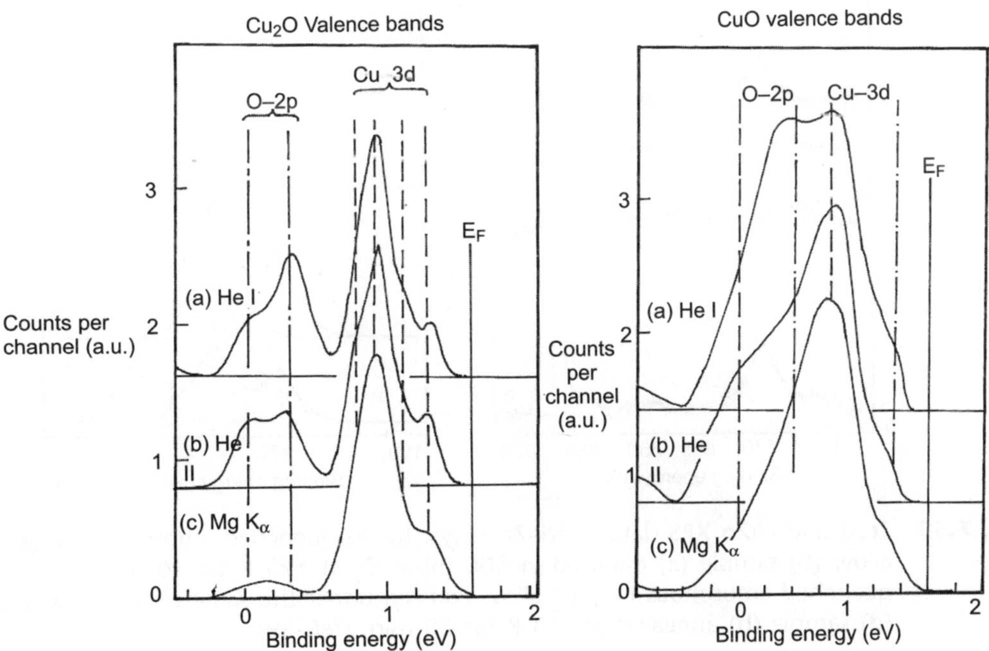

Fig. 7.52   Comparison of photo electron spectra of Cu_2O and CuO with He I, II and Mg K_α as the sources (ref. 58)

(c) XPS and UPS of Ni-Zr alloys

Amorphous alloys produced by the melt quenching method are promising precursors of oxide supported catalysts because of the easy preparation and the unique selectivity and reactivity in several reactions. The as-quenched amorphous alloys are usually inert to catalytic reactions, and special activation processes, such as oxidation, reduction or *in situ* catalytic reactions must be performed to generate the active sites. The pretreatments at crucial stages to activate the amorphous catalysts usually give rise to remarkable changes of the chemical composition and morphology, as well as of the crystalline structure of the alloy surface. Amorphous Ni–Zr alloys as typical amorphous catalysts show significant catalytic activities in the hydrogenation or dehydrogenation of olefins. It was found that the high activity was always obtained after cycles of oxidation in oxygen followed by reduction with hydrogen. The oxidation behavior of the amorphous Ni–Zr alloys with various alloy compositions, such as $Ni_{64}Zr_{36}$ and $Ni_{91}Zr_{9}$, has been studied extensively using XPS and UPS. Ni 2p and Zr 3d spectra under different environments are given in the Fig. 7.53.

The XPS studies on the oxidized amorphous $Ni_{64}Zr_{36}$ alloy revealed that the oxidation in air at elevated temperatures resulted in a significant enrichment of zirconium due to the formation of a zirconia layer on the surface. Similarly the oxidized sample followed by reduction enhances the Ni on the alloy surface. From the XPS we can understand that the oxidized sample is showing higher peak area for Zr than that of Ni.

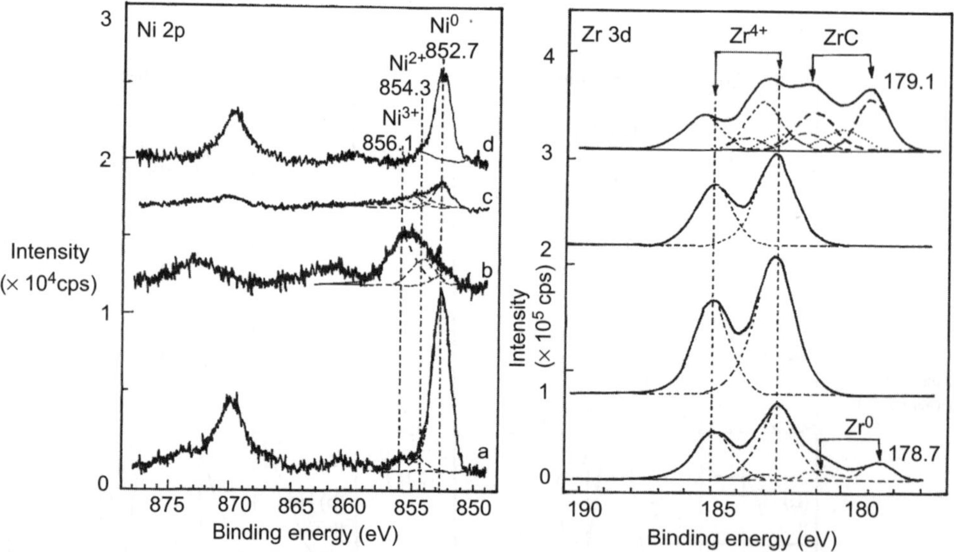

Fig. 7.53 Zr3d and Ni2p XPS data of Ni–Zr alloys. (a) As-quenched amorphous $Ni_{36}Zr_{64}$ alloy; (b) sample (a) oxidized in 500 mbar O_2 at 523 K for 30 min. ; (c) as-quenched amorphous $Ni_{24}Zr_{76}$ alloy oxidized under the same condition as (b); (d) sample (b) annealed at 773 K for 20 min. (ref. 59)

UPS data of the Ni$_{36}$Zr$_{64}$ alloy

The changes of the valence band structures of the amorphous Ni$_{36}$Zr$_{64}$ alloy due to different treatments have been studied by UPS. The spectra generated by using HeI (21.2 eV) as the exciting source are shown in Fig. 7.54. The valence spectrum of the as-quenched Ni$_{36}$Zr$_{64}$ exhibited a dominant peak at about 6 eV below the Fermi level (trace a). Similar peaks have been observed in UPS investigations of transition metal oxides, and have been attributed to emission from O 2p states interacting with the d-bands of metals. Combined with the corresponding XPS results, the main peak in trace 'a' can be assigned to surface ZrO$_2$. The shoulder centered at about 11 eV can be correlated with surface OH or C-related species, which are commonly present on untreated metal surfaces. Annealing to 773 K, however, gave rise to significant changes of the valence band spectrum, such as the emergence of the Fermi edge and a peak at about 3 eV below the Fermi level. Additionally, the intensity of the peak at 6 eV decreased strongly, as shown in Fig. 7.54. The appearance of the Fermi edge proves the metallic character of the annealed surface. Both the Fermi edge and the peak at 3 eV below the Fermi level can be assigned to ZrC at the surface. In the valence band spectrum of clean ZrC, a Fermi edge and a large structure distribution distributed between 0 and 5 eV were observed originating from hybridized Zr 4d and C 2p states.

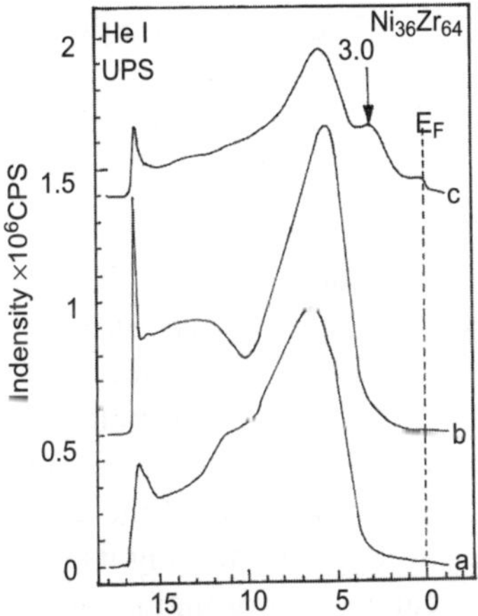

Fig. 7.54   (a) As-quenched amorphous alloy; (b) sample (a) oxidized in 500 mbar O$_2$ at 523 K for 30 min; (c) sample (b) annealed at 773 K for 20 min (ref. 59)

7.5.5 Angle Resolved Ultra Violet Photoelectron Spectroscopy

The UPS spectra of CO on Cu are similar to the spectra obtained on other transition metals, i.e., the peak at ~ 8 eV is assigned to overlapping emission from 5 σ and 1 π, while the peak at ~ 11 eV is due to emission from 4 σ level. A typical spectra for CO adsorbed on Cu (111) at 77 K are shown in Fig. 7.55. The shoulder at the higher binding energy side of the peak due to 4 σ is considered to be a shake up structure.

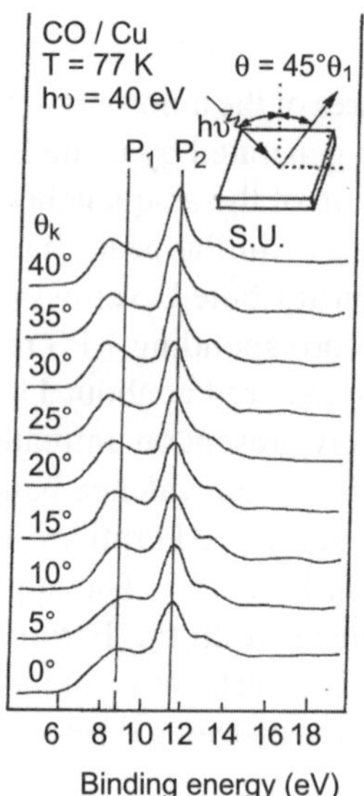

Fig. 7.55 UV photoelectron spectra in the angle –resolved mode for the adsorption of CO on Cu (111) at 77 K. The angular mode is shown at the top of the figure (ref. 60)

7.6 PERCEPTIONS

Wherever the properties of a solid surface are important, it is also important to have the means to measure those properties. The surface of solids plays an overriding part in a remarkably large number of processes, phenomena and materials of technological importance. These include not only catalysis but also corrosion, passivation, adhesion, tribology, metals, ceramics, micro electronics, composites, polymers, protective coatings, semiconductors, superconductors, interfaces and even grain boundaries. Photoelectron spectroscopy is one of the exciting and rapidly growing branches in solid state physics, chemistry and biology contributing its maximum for the study of these solid surfaces. Photoelectron spectroscopy can give information on the Fermi energies and the density of states in the valence bands of metals, alloys and semiconductors and to detect the chemisorption levels which are of vital importance in understanding the nature of chemisorption which is responsible for the surface reactions.

It is clear that the photo electron spectroscopy can be suitably exploited for studying a variety of processes in catalysis. These include (I) Identification of the active sites (II) The mode of generation of active sites (III) the pores present in active systems (IV) the mode of deactivation of a catalyst and the activation procedures required for the deactivated catalysts and so on. It should also be mentioned that many other properties of the catalyst solids, like their magnetic behaviour, electrical conduction, and extent of ionicity or covalency can also be deduced by suitable and appropriate photoelectron spectroscopic measurements. The examples of the systems given in this presentation are representative ones and they need not be the appropriate model system for the phenomenon discussed. The reader is referred to many other innumerable references that have been reported in literature for an extensive understanding of the phenomenon that can effectively studied by PES.

REFERENCES

This chapter was composed in collaboraton with Ms. C.M. JANET

1. K. Siegbahn, Nobel lecture; Physics (1981) 63.
2. K. Siegbahn, C. Nordling, A. Fahlman, H. Hamrin, J. Hedman, G. Johansson, Bergmark T., Karlsson S.E., Lindgren. J. and Lindberg B., Electron Spectroscopy for Chemical Analysis. Atomic, Molecular and Solid State Structure Studies by Means of Electron Spectroscopy (Uppsala: Almquist and Wiksells) (1967).
3. K. Siegbahn et *al. ESCA Applied to Free Molecules* (Amsterdam: North-Holland) (1969).
4. C. Nordling, E. Sokolowski and K. Siegbahn, *Phys. Rev.,* 105 (1957)1676–7.
5. D.W. Turner, C. Baker, A.D. Baker and C.R. Brundle, *Molecular Photoelectron Spectroscopy* (London: Wiley-Interscience) (1970).
6. D.W. Turner and M.I. Al-Joboury, *J. Chem. Phys.,* 37 (1962)3007–8.
7. Friedrich Reinert and Stefan Hufner, *New J. Phys.,* 7 (2005) 97.
8. J.C. Green, *Annual Review of Physical Chemistry*, 28 (1977)161-183.
9. W.E. Spicer and C.N. Berglund, *Phys. Rev. Lett.,* 12 (1964) 9–11.
10. C.N. Berglund and W.E. Spicer, *Phys. Rev.,* 136 (1964) 1030.
11. http://mason.gmu.edu/~jschreif/728/728-Surface%20Techniques.pdf
12. W.F. Krolikowski, and W.E. Spicer, *Phys. Rev. B,* 1 (2) (1970) 478.
13. W.E. Spicer, Phys. Rev. 112, (1958)114.
14. B.J. Mulder, *J. Phys. E: Sci. Instrum.* 10, (1977) 490-492.
15. D.E. Eastman, J.J. Donelon, N.C. Hien, and F.J. Himpsel, *Nucl. Instrum. Methods,* 172 (1980) 327.
16. D.E. Eastman, Phys. Rev. B 25 (1982) 7388.
17. D.E. Eastman and J.K. Cashion, *Phys. Rev. Lett.,* 27 (1971)1520.
18. D.E. Eastman and J.K. Cashion, *Phys. Rev. Lett.,* 24 (1970) 310.
19. B. Viswanathan, S. Sivasankar and A.V. Ramaswamy, Catalysis Principles and Applications, Chapter 10, Narosa Publishing House, New Delhi, (2002)163-183.
20. http://www.chem.qmw.ac.uk/surfaces/scc/scat5_3.htm
21. A.R. Chourasia and D.R. Chopra, Chapters 42-43, Frank Settle, Handbook of Instrumental Techniques for Analytical Chemistry, Prentice Hall PTR, Upper Saddle River, NJQ (1997) 791-827.
22. C.R. Brundle, J.F. Watts, J. Wolstenholene Ewing's Analytical Instrumentation Handbook, (3rd edition) edited by Jack cazes, Maral Dekker, New York (2005) 399 – 427.
23. augustus.scs.uiuc.edu/ nuzzogroup/PPT/XPS%20Class%2099.PPT
24. M.P. Seah, Practical Surface Analysis, Chapter 1, (Eds., D. Briggs and M.P. Seah, Wiley, New York, (1990)
25. http://www.aps.anl.gov/Facility/Technical_Publications/lsnotes/ls172/ls172.html
26. B.E. Koel, D.E. Peebles and J.M. White, *Surf. Sci.,* 125 (1983) 739.
27. John C. Riviere and Sverre Myhra, Handbook of Surface and Interface Analysis, methods for problem solving; Chapter 27, Marcel Dekker Ltd., (January 27, 1998) 851-888.
28. A. Frank, *Pico-The Omicron NanoTechnology Newsletter,* 7 (2003) 1.
29. S. Narayanan and B. Viswanathan, *J. Sci IInd Res.,* 48 (1989) 229 – 239.
30. D.M. Hercules, *Trends in Analyt Chem.,* 3 (1984)125.

31. Brinen J.S. and Schmitt J.L., *J. Catal.*, 45 (1976) 274.

32. J.S. Brinen and J.L. Schmitt, *J. Catal.*, 45 (1976) 274.

33. S. Scire, S. Minico, C. Crisafulli, C. Satriano and A. Pistone. *Appl. Catal. B: Environ.* 40 (2003) 43–49.

34. V. Cimpeanu, V. Parvulescu, V.I. Parvulescu, M. Capron, P. Grange, J.M. Thompson and C. Hardacre, *J.catal.* 235 (1) (2005) 184 – 194.

35. Jung Hwa Kang, Eun Woo Shin, Woo Jae Kim, Jae Duk Park, and Sang Heup Moon, *J. Catal.*, 208 (2002)310–320.

36. Y. Takasu, R. Unwin, B. Tesche, and A.M. Bradshaw, *Surf. Sci.* 77, (1978) 219

37. C. Shi, A.M. Zhu, X.F. Yang and C.T. Au, *Appl. Catal. A: Gen.*, 28 (2005) 83-90.

38. I. Olivelos, M.J.P. Zurita, C. Scott, M.R. Goldwasser, J. Goldwasser, S. Rondon, M. Houalla, D.M. Hercules, *J. Catal.*, 171 (1997) 485

39. W.H. Madhusudhan, S. Kollali, P.R. Sarode, M.S. Hegde, P. Ganguly and C.N.R. Rao, *Pramana*, 12 (1979) 317

40. Yu. Larichev, I. Prosvirin , D. Shlyapin, N. Shitova, P. Tsyrul'nikov, V. Bukhtiyarov, *Kinetics and Catalysis*, 46(4)(2005) 597-602 (6)

41. S. Narayanan and Uma. K, *J. Chem. Soc. Faraday Trans. I*, 81 (1985) 2733.

42. S. Narayanan, *J. Scient. Ind. Res;* 44 (1985) 319

43. S. Narayanan, Zeolites, 4 (1984) 231.

44. G. Ertl, R. Hierl, H. Knoezinger, N. Thiele & H. P. Urbach, *Appl. Surf. Sci.*, 5 (1980) 49.

45. R.D. Srivatsava, J. Onuferko, J.M. Schulz, G.A. Jones, K.N. Rai & R. Athappan, *IInd. Engg. Chem. Fundamn.*, 21 (1982) 457.

46. Ng, K.T. and D.M. Hercules, *J. Phys. Chem.*, 80 (1976) 2094.

47. P. Dufresne, E. Payen, J. Grimbolt and O.J.P. Bonnelle, *J. Phys. Chem.*, 85 (1981) 2344.

48. J.C. Vedrine, G. Hollinger and T.M. Duc, *J. Phys. Chem.*, 82 (1978) 1515.

49. N. Tsud, V. Johanek, I. Stara, K. Veltruska and V. Matolin, *Surf. Sci.* 467 (1-3) (2000) 169-176.

50. B. Viswanathan, M.J. Omana and T.K. Varadarajan, *Cat. Lett.* 3 (1989) 217–222.

51. V.A.M. Brabers, F.M. Van Setten and P.S.A. Knapen, *J. Solid State Chem.*, 49 (1983) 93.

52. R. Vetrivel, K. Seshan, K.R. Krishnamurthy and T.S.R. Prasada Rao, *Bull. Materials Sci.*, 9 (1987) 75.

53. B.A. Sexton, *Materials Forum*, 10 (24) (1987) 134.

54. Ch. Papadopoulou, J. Vakros, H.K. Matralis, Ch. Kordulis and A. Lycourghiotis, *J Colloid Interface Sci.*, 261-1(2003) 146-153.

55. D.E. Eastman and M.I. Nathan, *Phys. Today*, 28 (1975) 44-51.

56. C.R. Brundle, P.S. Bagus, D. Menzel, and K. Hermann K, *Phys. Rev.*, B 24 (1981) 7041.

57. J. Michael Hollas, Modern spectroscopy, 4th edition, John Wiley & Sons, New York, January 2, (2004).

58. S. Hufner, Photoelectron Spectroscopy—Principles and Applications, Springer-Verlag, Berlin, (2003).

59. Z. Song, X. Bao, U. Wild· M. Muhler and G. Ertl, *Appl. Surf. Sci.*, 134 (1-4) (1998) 31-38.

60. Shin-ichi Ishi and Yuichi Ohno and B. Viswanathan, *J. Sci. IInd. Res.*, 46 (1987) 541-567.

X-ray Diffraction—Basics and Applications for Catalysts Characterization

8.1 INTRODUCTION

X-rays were discovered in 1895 by the German physicist W.C. Röntgen (for which he was awarded the first Nobel Prize in Physics in 1901) and were so named because their nature was unknown at that time. Unlike ordinary light, these rays are invisible, but they traveled in straight lines and affected photographic film in the same way as light. This property immediately attracted (despite the fact that the property of the materials were not known!!) initially physicians (X-ray radiography) and later on by engineers to unravel fractures in bones of human beings and identification of cracks and defects in engineering materials like metal castings respectively. It was not until 1912, the remarkable invention by Laue on X-ray *diffraction* by crystals, proved two important findings namely the wave nature of X-rays and possibility of finding fine structures of materials (in the order of *nanometers*). Characteristic secondary radiation emission upon excitation with a high energy X-ray source led to the birth of X-ray fluorescence mainly used to determine the amounts of elements in materials. Pictorially, the uses of X-rays are given in Fig. 8.1. This article essentially dwells on certain fundamentals of X-ray diffraction including some historical milestone, the techniques employed comprising those of current times and applications of this technique in varied facets of catalysis science and technology.

8.2 THEORY OF DIFFRACTION

X-rays and Their Generation

X-rays are electromagnetic radiation whose wavelength is of the order of 10^{-10} m (roughly the size of an atom; $1\text{Å} = 10^{-10}$ m) falls between ultraviolet and gamma rays in the electromagnetic spectrum (Table 8.1). X-rays that are used in diffraction measurements, have wavelengths of the order of 0.5 to 2.5 Å. They are produced when any electrically charged particles (usually electrons) of sufficient kinetic energy rapidly decelerate through an impact on a target (usually metal). When the rays coming from the target are analyzed, they are found to consist of a mixture of different wavelengths and the variation of intensity with wavelength is found to depend on the tube voltage. When the tube voltage is raised, the intensity of all wavelengths increases and the position of the all wavelengths shift to shorter wavelengths. The radiation represented by such curves is called polychromatic, continuous or white radiation. However, when the voltage on an X-ray tube is raised above a certain critical value, characteristic of the target metal, sharp intensity maxima appear at certain wavelengths superimposed on the continuous spectrum. Since they are narrow (of the order of 0.001Å) and wavelengths are characteristic of the target metal used, they are called characteristic lines or nearly monochromatic lines. Such characteristic lines make a great deal of X-ray diffraction possible, because most of the diffraction experiments require the use of monochromatic or approximately monochromatic radiation.

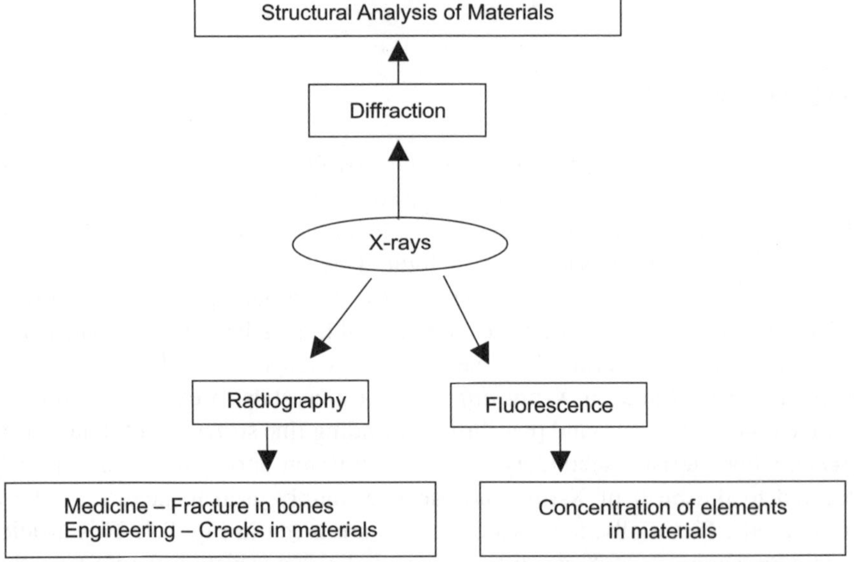

Fig. 8.1 Pictorial applications of X-rays

Table 8.1 The electromagnetic spectrum

Region	Wavelength		Frequency (Hz)	Energy (eV)
	(in Å)	(in cm)		
Radio	$>10^9$	>10	$<3\times10^9$	$<10^{-5}$
Microwave	10^9-10^6	$10-0.01$	$3\times10^9-3\times10^{12}$	10^{-5} to 0.01
Infrared	10^6-7000	$0.01-7\times10^{-5}$	$3\times10^{12}-4.3\times10^{14}$	0.01 to 2
Visible	$7000-4000$	$7\times10^{-5}-4\times10^{-5}$	$4.3\times10^{14}-7.5\times10^{14}$	$2-3$
Ultraviolet	$4000-10$	$4\times10^{-5}-10^{-7}$	$7.5\times10^{14}-3\times10^{17}$	$3-10^3$
X-rays	$10-0.1$	$10^{-7}-10^{-9}$	$3\times10^{17}-3\times10^{19}$	10^3-10^5
Gamma rays	<0.1	$<10^{-12}$	$>3\times10^{19}$	$>10^5$

Practically all X-ray tubes contain two electrodes, an anode (which is the target metal) maintained at ground potential and a cathode, maintained at high negative potential, normally of the order of 30,000 to 50,000V. Filament tubes which were invented by Coolidge in 1913 consist of an evacuated glass envelop wherein tungsten filament is heated by a filament current of about 3 amp. and the emitted electrons are rapidly drawn to the target due to the high voltage gradient across the tube. Surrounding the filament is a small metal cap which is also maintained at high voltage as the filament; it therefore repels the electrons and tends to focus them into the narrow region of the target, called the focal spot. X-rays emitted from the focal spot in all directions and escape from the tube through two or more windows in the tube housing. Since these windows must be vacuum tight and yet highly transparent to X-rays, they are usually made of beryllium. Many diffraction experiments require radiation which is as closely monochromatic as possible. However, the beam from an X-ray tube operated at a voltage above V_k (critical voltage) contains not only K_α line but also the weaker K_β line and the continuous spectrum. The intensity of the undesirable components can be decreased relative to the intensity of the K_α line by passing the beam through a filter made of a material whose K absorption edge lies between K_α and K_β wavelength of the target metal. Such a material will have an atomic number one less than that of the target metal, for metals with atomic number near 30. For example, for copper target, nickel is used as filter which decreases the intensity of K_β/K_α from about 1/9 in the incident beam to about 1/500 in the transmitted beam. A representative figure of modern X-ray tubes is given in Fig. 8.2.

A brief knowledge on crystal system is essential and hence the same is dealt with first.

8.2.1 Crystal Lattice

A crystal may be defined as a solid composed of atoms, ions or molecules arranged in a pattern periodic in three dimensions. Crystals, different from the other forms of matter say, primarily liquids and gases which lack periodicity. Many solids are crystalline may be in the form of single crystals or they consist of many contiguous crystals known as polycrystalline. However, not all solids are crystalline, some are amorphous, for example, glass, which does not have any regular internal arrangement. Scientifically, no essential difference exists between an amorphous solid and a liquid. While thinking about crystals, it is essential to consider or focus on the geometry of the periodic arrays. The crystal may then be represented as a lattice, which is a three dimensional array of points (termed as lattice points) each of which have identical surroundings. In defining a lattice with three non-coplanar lattice vectors, unit cells of various shapes can result, depending on the length and orientation of the vectors. For example, a, b, c are of equal length and at right angles to one another,

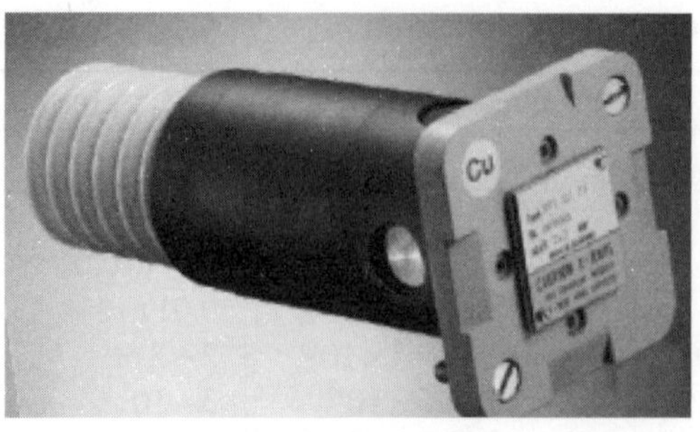 

Fig. 8.2 Modern ceramic X-ray tube used in diffractometers

or a = b = c and $\alpha = \beta = \gamma = 90°$, the unit cell is cubic. Given special values to the axial lengths and angles, produces unit cells of seven different kinds, which are sufficient to include all the possible point lattices which are commonly known as crystal systems, listed in Table 8.2.

Seven different point lattices can be obtained simply by putting points at the corners of the unit cells of the seven crystal systems. However, there are other arrangements of points which fulfill the requirements of a point lattice, namely, that each lattice point have identical surroundings. The French crystallographer Bravais worked on this and demonstrated that there are fourteen possible point lattices and named as Bravais lattice, illustrated in Fig. 8.3. There are excellent books given at the end of this chapter that provide an in depth understanding of crystal systems, as crystallography itself is a vast subject (in fact it is older than X-rays). With this simple knowledge about crystal systems we shall now proceed to the concept of diffraction.

Table 8.2 Seven crystal systems and Bravais lattices

System	Axial lengths and angles	Bravais lattice	Lattice Symbol
Cubic	Three equal axes at right angles $a = b = c$, $\alpha = \beta = \gamma = 90°$	Simple Body-centered Face-centered	p I F
Tetragonal	Three axes at right angles, two equal $a = b \neq c$, $\alpha = \beta = \gamma = 90°$	Simple Body-centered	P I
Orthorhombic	Three unequal axes at right angles $a \neq b \neq c$, $\alpha = \beta = \gamma = 90°$	Simple Body-centered Base-centered Face-centered	P I C F
Rhombohedral[*]	Three equal axes, equally inclined $a = b = c$, $\alpha = \beta = \gamma \neq 90°$	Simple	R
Hexagonal	Two equal coplanar axes at 120°, third axis at right angles $a = b \neq c$, $\alpha = \beta = 90°$ ($\gamma = 120°$)	Simple	P
Monoclinic	Three unequal axes, one pair not at right angles $a \neq b \neq c$, $\alpha = \gamma = 90° \neq \beta$	Simple Base-centered	P C
Triclinic	Three unequal axes, unequally inclined and none at right angles $a \neq b \neq c$, $\alpha \neq \beta \neq \gamma \neq 90°$	Simple	P

[*]Also called trigonal

Diffraction

Diffraction is essentially the interaction of X-rays and crystal systems. Studies on X-rays proved that they can be polarized, but could not be refracted led to controversies over wave-particle issue in the early 20th century. The concept of diffraction (although not diffraction by X-rays) was known, however, for visible light wherein whenever a wave motion encountered a set of regularly spaced scattering objects (e.g., ruled grating), provided that the wavelength of the wave motion was the same order of magnitude as the repeat distance between the scattering centers, diffraction occurs. Such was the knowledge in 1912 when the German physicist von Laue took up the problem. von Laue reasoned that, *if* crystals are composed of regularly spaced atoms which might act as scattering centers for X-rays and *if* X-rays were electromagnetic waves of wavelengths about equal to the interatomic distance in crystals, then it should be possible to diffract X-rays by means of crystals. He devised an experiment in which X-rays were allowed into a lead box containing a crystal of copper sulfate, with sensitive film behind the crystal. In the second attempt, when the films were developed there was a large central point from the incident X-rays, but also many smaller points in a regular pattern. These could only be due to diffraction of incident beam and interference of many beams. By this, he proved that X-rays were waves of light with very small wavelengths and published his results in 1912 for which he was awarded the Nobel prize in physics for the year 1914.

Diffraction is due essentially to the existence of certain phase relations between two or more waves. A diffracted beam may be defined as a beam composed of a large number of scattered rays mutually reinforcing one other. Diffraction, is essentially a scattering phenomenon, wherein the atoms of the crystal lattice scatter incident X-rays in all directions and at some of these directions the scattered beams will be completely in phase and hence reinforce each other to form a diffracted beam. Mathematically W.L. Bragg formulated (with the guidance of his father W.H. Bragg) the conditions at which such reinforcement of waves occurs and defined a famous law known as Bragg's law which can be written as

$$n\lambda = 2d\ sin\theta$$

where λ is the wavelength of the incident X-ray beam, θ is the angle of reflection and d is the interplanar spacing.

Based on this law, he solved the structures of NaCl, KCl, KBr and KI (all of them have NaCl structure), which were the first complete crystal-structure determination ever made in the history (Sir William Henry Bragg, Sir William Lawrence Bragg shared the Nobel prize in 1915). Experimentally, Bragg's law can be applied in two ways. By using X-rays of known wavelength and measuring θ, the spacing d of various planes in a crystal is determined: this is *structure analysis*. Alternatively, a crystal with planes of known spacing d can be used to measure θ and thus determined the wavelength λ of the radiation used and this is *x-ray spectroscopy*. The scope of this chapter is essentially on the former subject namely "structure analysis" and this can be done (in order to satisfy Bragg's law) by continuously varying either λ or θ during the experiment. In general three diffraction methods were followed namely:

Method	λ	θ
Laue	Variable	Fixed
Rotating-crystal	Fixed	Variable (in part)
Powder	Fixed	Variable

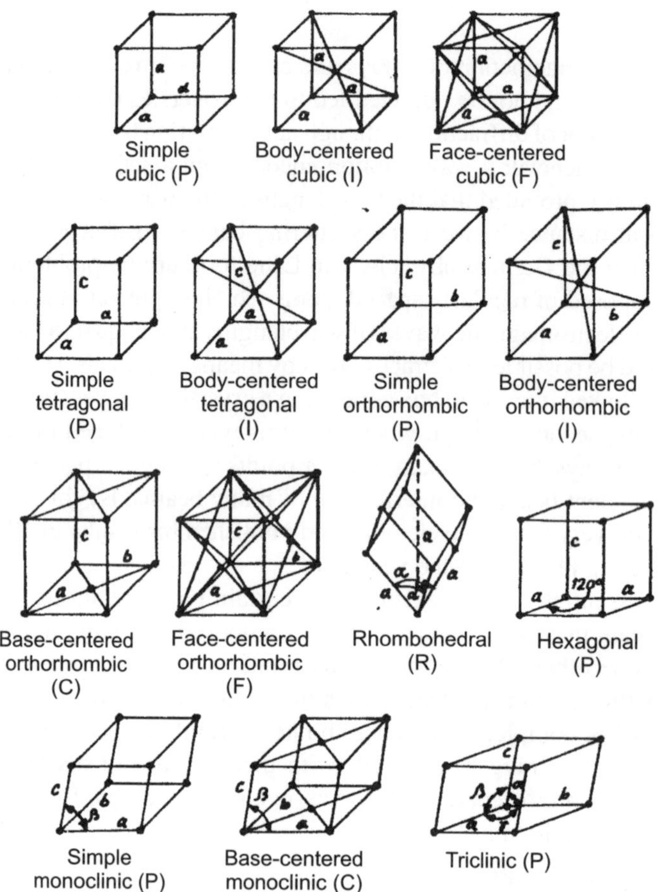

Fig. 8.3 The fourteen Bravais lattices (taken from Elements of X-ray diffraction, Cullity and Stock, Prentice Hall, 2001, p. 45)

Now a days, *powder* X-ray diffraction is widely practiced for numerous applications, although diffraction of thin films and gels are also reported for specific applications. Having understood the fundamentals of diffraction and the criteria/conditions essential for the diffraction to occur, let us discuss briefly about the instrumentation of X-ray diffraction.

8.2.2 Diffractometer: Features and Advances

The method of X-ray powder diffraction was devised independently in 1915 by Debye and Scherrer, in Germany and in 1916 by Hull, in United States, and both of them have used photographic films to record the angles and intensities of the diffracted beam. There was a quite a long gap (except for the few home made diffractometers) of nearly 25 years for a commercial diffractometer (in 1940s) to come into the market. But they were rapidly growing thanks to the development of material science (as it yielded excellent information on the structures of materials) and certain advantages offered by the commercial ones over the conventional film techniques (the era of film to paper).

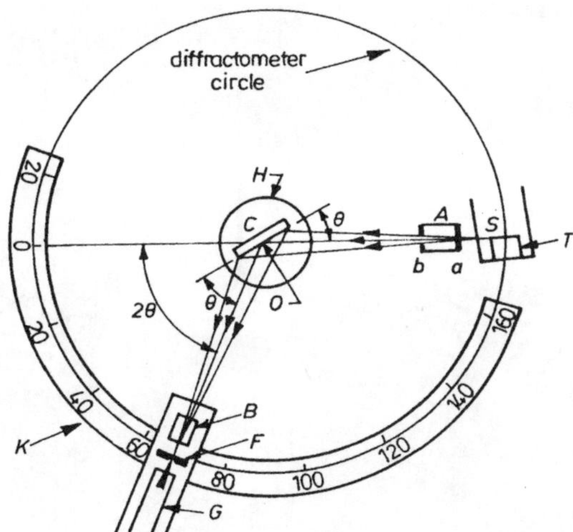

Fig. 8.4 Schematic diagram of X-ray diffractometer (taken from Elements of X-ray diffraction, Cullity and Stock, Prentice Hall, 2001, p.187)

Essentially even today the diffractometer is designed somewhat similar to what Hull and Debye-Scherrer have done in early 20th century, however, with lots of advancements, thanks to electronics and instrumentation. Essentially, monochromatic radiation is used and the X-ray detector is placed on the circumference of the circle centered on the powder specimen. The essential features of a diffractometer are shown in Fig. 8.4. A powder specimen C, in the form of a flat plate, is supported on a table H, which can be rotated about an axis O perpendicular to the plane of drawing The X-ray source S, in the line focal point on the target T of the X-ray tube; S is also normal to the plane of drawing and therefore parallel to the diffractometer axis O. X-rays diverge from this source and diffracted by the specimen to form a convergent diffracted beam which comes to a focus at the slit F and then enters the detector G. A and B are special slits which define and collimate the incident and diffracted beam. The monochromator or filter is usually placed in a special holder in the diffracted, rather than the incident beam; a monochromator or filter in the diffracted beam not only serves its primary function (suppression of K_β radiation) but also decreased background radiation originating in the specimen. The receiving slits and detector are supported on a carriage E, which may be rotated about the axis O and whose angular position 2θ may be read on the graduated scale K. The supports E and H are mechanically coupled so that a rotation of detector through 2x degrees is automatically accompanied by the rotation of the specimen through x degrees (commonly known as Bragg-Brentano geometry). A strong advancement has happened in the last few decades owing to the development of electronics and instrumentation, wherein the control of the diffractometer is done through a computer (including mechanical movements) and the data are collected, processed and analyzed by using various commercial software. Fig. 8.5 and Fig. 8.6 shows the recent diffractometers offered by major commercial vendors.

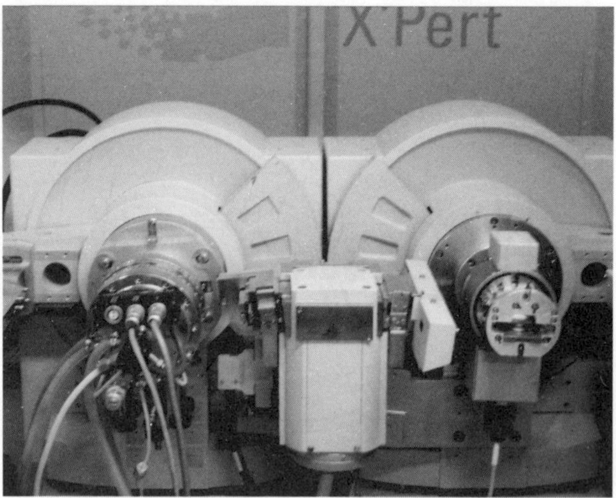

Fig. 8.5 Philips X'Pert MPD System (Bragg-Brentano Geometry at CSMCRI, Bhavnagar)

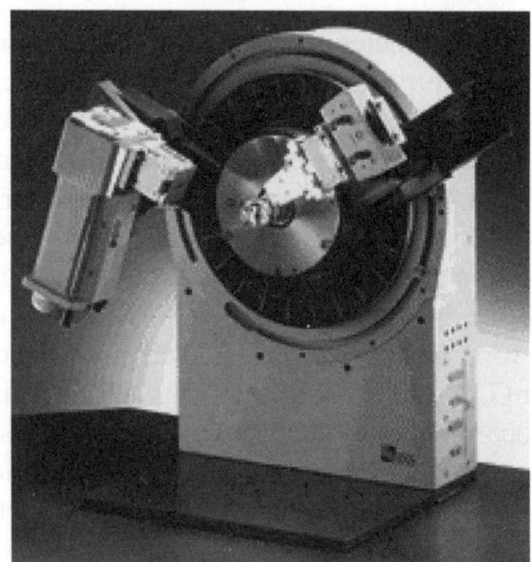

Fig. 8.6 Bruker AXS X-ray Diffractometer (Parallel Beam Optics)

8.2.3 Applications of Powder X-ray Diffraction (PXRD)

PXRD has a wide range of applications in various fields of science and engineering namely in geology, material science, catalysis, environmental science and more recently and extensively in pharmaceutical industry. Each field has its own nature of application and in this text focus will be mainly emphasized on catalytic applications. The applications are represented in the form of case studies which will make a better understanding of exploiting this tool for deriving varied information which are indispensable for science and technology of catalysis.

8.2.4 Phase Identification

Case Study 1: Fischer-Tropsch Reaction

Fischer-Tropsch reaction is one of the important reactions for the production of variety of alcohols from CO and H_2 using metal supported oxide catalysts. One of the systems widely studied in literature is Fe supported MnO [1]. Fig. 8.7 shows the PXRD patterns of the catalyst studied before and after the reaction. For the sake of comparison and to draw conclusions, reference pattern of MnO is also included in the figure. The PXRD pattern of reduced Fe/MnO catalyst (Fig. 8.7 B) showed the presence of a peak at $2\theta = 57°$, a characteristic peak for body centered cubic iron. But if one were to look at the MnO reflections, two important observations could be noted wherein the peaks have become broad and also shifted to higher angles (compared to reference MnO). The shift to higher angles (in other words, lower d values) suggest partial incorporation of Fe^{2+} in MnO matrix whose ionic radius is smaller than that of Mn^{2+} ($Mn^{2+} = 0.81$Å and $Fe^{2+} = 0.71$Å). The broadening of peaks could be reasoned for smaller particle size (Scherrer equation), the details of them are discussed in another case study. Further information emerged while studying the spent catalyst (catalyst after the reaction; Fig. 8.7 C). It showed the absence of the peak at $2\theta = 57°$ suggesting the absence of iron while less intense reflections were noted at $2\theta = 47, 49, 55$ and $56.5°$. A search-match procedure (either using Hannavalt procedure or computer-based database interfaced search-match programs)

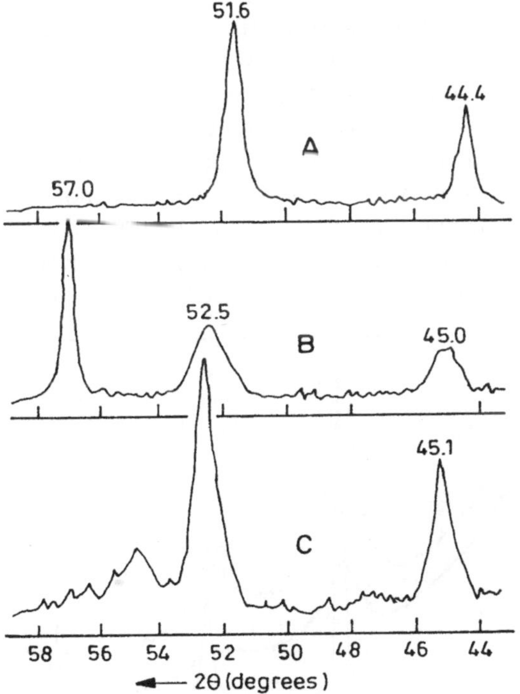

Fig. 8.7 X-ray diffraction in the identification of phases in manganese-promoted iron Fischer Tropsch catalyst, (A) MnO (reference), (B) reduced Fe-MnO catalyst and (C) catalyst after CO hydrogenation (reproduced from ref. 1)

showed the presence of iron carbide. This indeed gave clues on the activation-deactivation behavior of the catalyst, resulted in the development of better catalysts. Such conclusions were also complimented and supported by Mossbauer spectroscopy.

Case Study 2: Crystallite Size and Line Broadening: Supported Palladium Catalysts for Hydrogenation

One of the essential criterion of powder X-ray diffraction is the material under test should possess sufficient long range ordering to give adequate constructive interference which is sufficient enough to be detected by detectors to derive appropriate conclusions. However, scientists are also innovative to use this limitation for their advantage in obtaining some information which is critical/essential for certain applications. The advantage of this limitation is that the width (or the shape of the peak) of diffraction peaks carries information on the dimensions of the reflecting planes. A classic example is palladium supported on silica for hydrogenation reactions [2]. Fig. 8.8 shows PXRD pattern of palladium whose reflections namely (111) and (200) are quite sharp and narrow which occurred from perfect crystals. However for crystallite sizes below 100 nm, line broadening occurs due to incomplete destructive interference in scattering directions where X-rays are out of phase. Fig. 8.8 shows diffraction patterns of two supported Pd catalysts with different loadings of Pd (1.1 and 2.4 wt.%), exhibited broad reflections. It is quite clear that the intensity of the palladium reflections directly depended on the concentration of palladium. Applying Scherrer equation ($t = 0.89\lambda/\beta \cos\theta$, where t is crystallite size, β is FWHM and θ is diffraction angle), the average diameters calculated were 42 and 25 Å for catalyst with 2.4 and 1.1% of Pd, respectively. In other words, lower the loading, lower the crystallite size, better the dispersion and better the intrinsic activity. It must however be cautioned here that X-ray broadening provides a quick but not always reliable values of crystallite size. The sizes are calculated through statistical evaluation weighted by the volume of the particles and not by the number or by surface area, as would be more meaningful for catalysis, as it is essentially a surface phenomenon. One should also be careful to consider instrumental broadening in the 2θ region of measurement and internal strain of the particles in contributing to such broadening while deriving at these results, part of these are covered in another case study mentioned below.

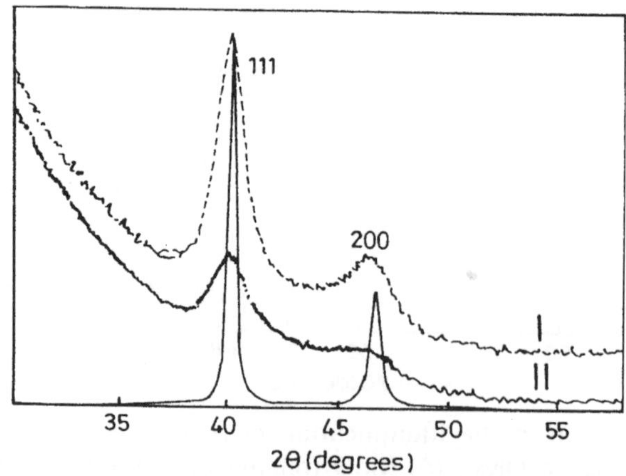

Fig. 8.8 XRD patterns of silica supported palladium catalysts (I: 2.4wt% Pd; II: 1.1 wt% Pd), and of a palladium reference sample (reproduced from ref. 2)

8.2.5 Lamellar Structures for Solid Acid and Base Catalysis

Powder diffractometry is an excellent tool to study the samples with layered structure. In general, such layered materials exhibit preferred orientation (i.e., (00l) planes of the sample are parallel to the reference plane of the sample holder) so that the intensity of the basal reflections namely (001), (002) and (00l) are significantly increased. In addition, intercalation of any ions (whether it is positively charged species in the case of montmorillonite type materials or negatively charged species in the case of hydrotalcite-like materials) in the interlayer of these materials lead to significant shift in the basal reflections depending on the size, charge and orientation of the ions thereby ascertain the occurrence, degree and nature of such intercalation process. This is evidenced by couple of case studies listed below:

Case Study 3: Intercalation of Polyoxocations in Rectorite as FCC Catalysts

Rectorite, when pillared with polyoxocations of Al or Zr, can form microporous solids with good thermal and hydrothermal stabilities comparable to that of zeolites based materials which are used in fluidized catalytic cracking (FCC). Structurally, rectorite is a mixed-layer (Figs. 8.9 - 8.11) clay consisting of regular intrastratification of low-charge density montmorillonite-like layers and high-charge density (nonexpendable) mica like layers. The charge compensating cations in the swelling layers can be easily replaced by large oxocations such as $[Al_{13}O_4(OH)_{24}(H_2O)_{12}]^{7+}$ [3]. Once they are introduced, layers of the montmorillonite prop apart. This can be clearly seen from powder X-ray diffraction, given in Fig. 8.9-I. When such intercalation occurs, the d_{001} spacing expands from 24.1 to 28.7 Å and the resulting clay surface area has moderately increased (180 m²/g). After drying and calcination at 400°C for 2 h in air, partial dehydroxylation of Al_{13}-pillars occurs, which reflect in the reduction of d_{001} spacing to lower values (28.0 Å). Further calcination at 760°C in 100% steam does not significantly change the diffraction profile (although surface area reduced to 147 m²/g) suggesting the stability of such pillared structure even after such severe hydrothermal treatment.

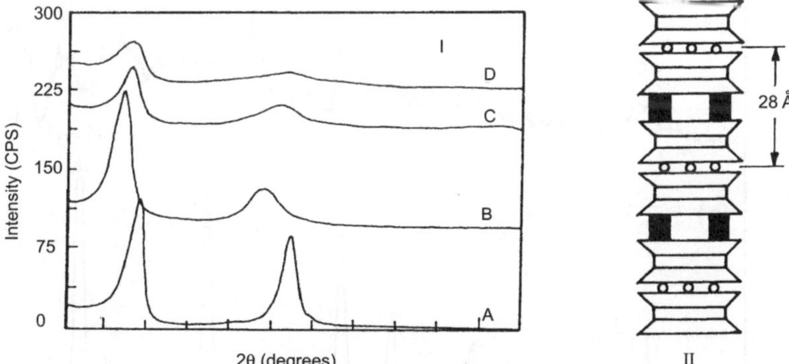

Fig. 8.9 X-ray diffractograms of Mg-rectorite (I) (A) Before pillaring (B) After pillaring of alumina clusters (c) Calcination of sample B at 540°C in air and steaming at 760°C for 5h in the presence of 0.5% V and (d) 1.0% V. and (II) Schematic representation of rectorite (reproduced from ref. 3)

In other words, PXRD has been a tool of excellence to find the intercalation process as well as provide evidence for the hydrothermal stability of such layered catalysts. A complimentary case study in the case of anionic clays is given on next page.

Case Study 4: Intercalation of Metal Complexes in Hydrotalcites for Selective Oxidation

Incorporation of transition metal complexes into the cavities and/or layers of the host materials such as zeolites and clays is a subject of recent interest owing to their advantageous catalytic properties in selective oxidation reactions [4].

Among various host lattices, hydrotalcite-like lattice offers possibilities for incorporation of anionic complexes through intercalation [5]. Such confined complexes are expected to be advantageous over their homogeneous counterparts owing to site isolation and provide easy separation and regeneration. Fig. 8.10 shows the PXRD pattern of $MgAlCO_3$-HT which is a naturally occurring hydrotalcite and MgAl-[CuPcTs]-HT where CuPcTs = Copper(II)phthalocyanine tetrasulfonate. In general for a carbonate containing hydrotalcite, the basal reflection namely (003) appears around d = 7.8 Å. Intercalation of metallophthalocyanine complex shifted this reflection to lower angles thereby d_{003} has a value around 13.0 Å. However, intercalation of such robust anion disturbs the long range ordering of the sheets thereby affecting the crystallinity of the material. In addition, based on the molecular dimension of the anion and after taking into consideration of the layer thickness (4.8 Å), one can say that the anion is neither oriented parallel nor perpendicular to the layers but in an inclined orientation. Oxidation of cyclohexanol to cyclohexanone was carried out over this intercalated complex. Comparison of the normalized activity between the neat complex over the intercalated complex revealed a five times better activity for the latter catalyst suggesting its superiority. Similar line of thought is extended for intercalating polyoxovanadate anion for oxidative dehydrogenation of propane, osmate for asymmetric dehydroxylation of various organic substrates and tungstate for bromide oxidation [6-8].

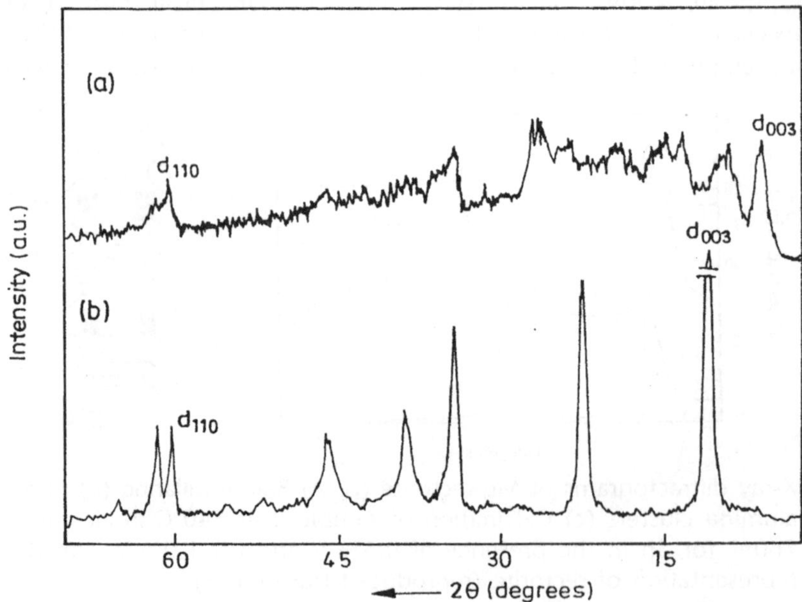

Fig. 8.10 PXRD patterns of (a) MgAl[CuPcTs]-HT and (b) $MgAlCO_3$-HT where CuPcTs is copper phthalocyanine tetrasulfonate (reproduced from ref. 5)

8.2.6 Solid Solutions and Their Synergism in Catalysis

Incorporation of Transition Metal Ions in Zeolite Framework

Incorporation of metal ions in a matrix (whether oxide or aluminosilicate) thereby resulting in solid solutions or isomorphous substitution is one of the art heterogeneous catalyst scientists have mastered over the period of years to make step-change in catalytic behavior. It is also proved that a simple physical mixture of two components doesn't result in such a phenomenon while structural admixture sometimes results in such augmented behavior. In addition, the extent of such incorporation could also be verified through diffraction from the variations of unit cell parameter (if Vegard's law is obeyed) thereby giving clues on the range of such solid solutions/incorporation possibilities. The following two case studies, one on the isomorphous substitution of transition metal ions in zeolites and the other on the oxide solid-solution could evidently propel such a phenomenon:

Case Study 5: Isomorphous Substitution of Si^{4+} by Zr^{4+} for Selective Oxidation

Isomorphous substitution of Si^{4+} by metal ions such as Al^{3+}, Zr^{4+} etc. is a subject of continual interest owing to the varied redox properties possessed by the resulting material. Of the transition metal ions studied, Zr^{4+} is one of the interesting metal ions owing to its bifunctional behavior. This case study reveals how such isomorphous substitution of Zr^{4+} in MEL topology of silicalite (referred here as Sil-2) lattice is revealed through powder X-ray diffraction and how such a material has been exploited for catalytic reaction [9]. Incorporation of Zr is done through conventional hydrothermal synthesis.

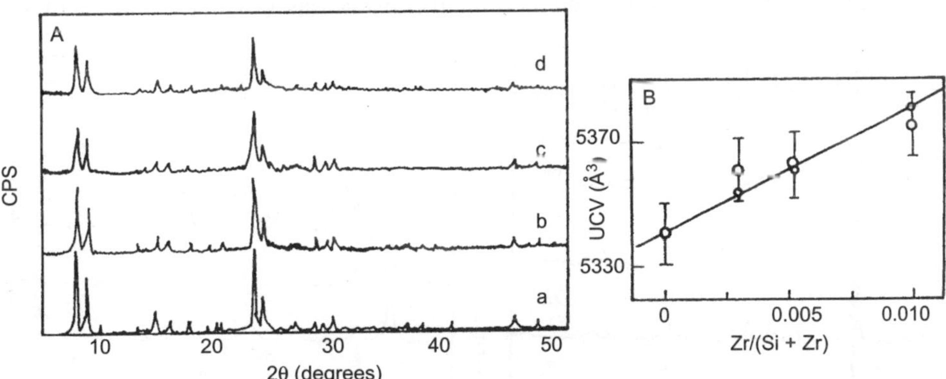

Fig. 8.11 X-ray diffraction profiles of calcined silicalite-2 (MEL) (curve a) and Zr-Sil-2 samples prepared by hydrothermal with Si/Zr mole ratios of 300, 200 and 100 (curve b to d respectively). (B) Increase in unit cell volume with Zr content in the samples (reproduced from ref. 9)

The PXRD profiles (Fig. 8.11-A) of calcined Zr-Sil-2 are similar to Sil-2 (without any impurity phases) except for subtle variations in the crystallinity. However, a closer look at the interplanar spacing revealed a higher value for Zr-incorporated sample. This can be attributed to higher ionic radius of Zr^{4+} (0.59 Å) over that of Si^{4+} (0.26 Å) in tetrahedral co-ordination. Further, variation of unit cell dimension with Zr^{4+} loading (up to certain extent) showed (Fig. 8.11-B) reasonable linear correlation (correlation co-efficient is better than 0.9) suggesting the compliance of Vegard's law.

Such incorporation is also verified by other physicochemical means. Hydroxylation of phenol showed a pronounced activity using H_2O_2 as oxidant wherein both neat silicalite-2 as well Zr-impregnated silicate-2 showed no activity under similar reaction conditions suggesting the inherent activity of Zr^{4+} present in the lattice.

Case Study 6: Mixed Metal Perovskite-type Oxides for deNOx Catalysis

Perovskite-type oxides with the general formula ABO_3 are one of the inexpensive substitutes for noble metal catalysts for CO oxidation, NO reduction or oxidation of hydrocarbons and hence visualized as the best substitute catalyst for automobile exhaust. The ideal perovskite structure belongs to the cubic space group P$m3m$. If the A cation is replaced by a smaller/bigger one, BO_6 octahedra would rotate about the cubic crystallographic axis to influence the structural stress leading change in the B-O-B bond angle and the coordination number of A cations [10]. Thus, by replacing part of A or B sites with A' or B' ions to form $A_xA'_{1-x}B_yB'_{1-y}O_{3\pm\lambda}$ structure, it is possible to create oxygen vacancies in the perovskite oxides and exhibit higher oxidation ability. Such an effort is made in substituting Pr in $PrMnO_3$ by strontium or cerium and how XRD assisted in design of such solid solutions will be evidenced in this case study. The samples were synthesized by sol-gel method and the final oxides were obtained by calcination at 750°C for 3 h.

The X-ray diffraction patterns of the Sr- or Ce-doped $PrMnO_3$ samples are given in Fig. 8.12 (a and b respectively). The main diffraction peaks of $PrMnO_3$, $SrMnO_{3-\lambda}$ and CeO_2 are included for comparison. For Sr-doped samples, it is possible to obtain approximately single perovskite type oxides after calcination up to x = 0.5 while above 0.5, peaks of other phases such as $SrMnO_{3-\lambda}$, $SrCO_3$ and Mn_3O_4 are evidently found and their intensities increased with an increase in Sr doping.

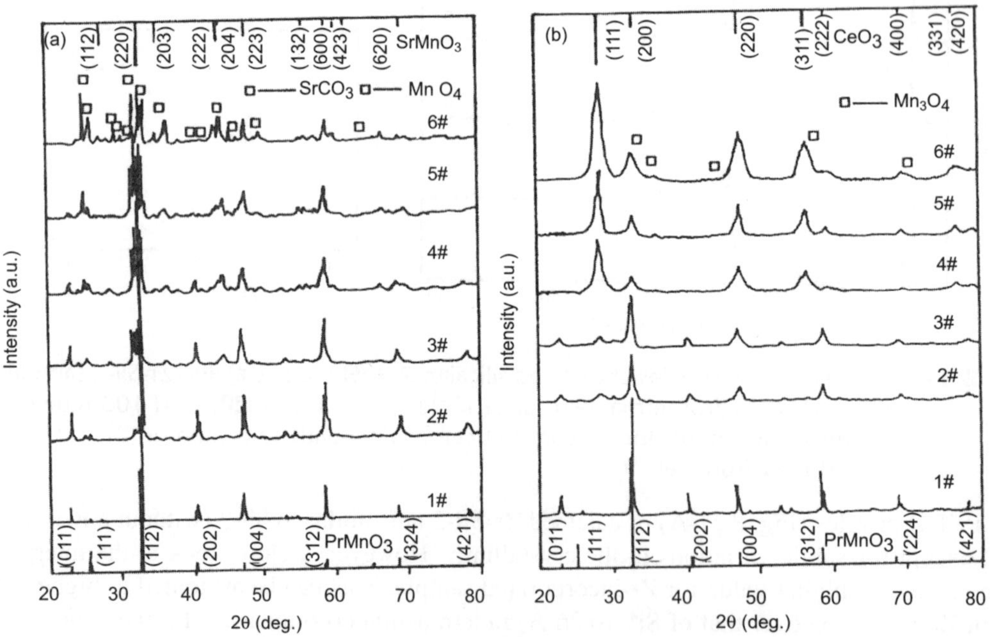

Fig. 8.12 XRD patterns of (a) Sr- or (b) Ce-doped $PrMnO_3$ where Pr:Sr/Ce = 1-x where x = 0, 0.2, 0.4, 0.6, 0.8, 1.0, referred to as 1-6 (reproduced from ref. 10)

The presence of $SrCO_3$ suggests that calcination temperature for formation of perovskite need to be increased. At higher values of x, the dominant peaks of perovskite were $SrMnO_{3-\lambda}$. In the case of cerium substitution, ceria phase was detected when x > 0.6 and less intense perovskite phase was detected.

Thus one can infer the limit at which substitution can be entertained in solid solution without forming discrete or undesired phase through PXRD. Such results are helpful in designing tailored catalysts. The solid solution exhibited enhanced catalytic activity for NO reduction when compared to pure $PrMnO_3$. Between Sr and Ce, the latter showed higher activity as it enhances redox property of the catalyst based on sorption/desorption cycles of lattice oxygen.

Case Study 7: Hydrotalcite-like Materials for Solid Base Catalysis

Hydrotalcite (HT) is a *layered material* consisting of alternating mixed metal hydroxides sheets separated by interlayer anions and water molecules. A continuing interest is emerging on the utility of these materials in solid base catalysis and in selective oxidation [11-13].

Structurally, it can be described as a close packing of hydroxyl anions in which cations occupy alternate layers of octahedral sites and such M-OH octahedral are edge shared to form infinite sheets which are stacked over one another and for electroneutrality anions occupy in the interlayer positions. In general these materials are synthesized through coprecipitation using metal nitrates using alkali and diversified combinations of metal ions have been introduced in HT-like lattice by different research groups across the world. However, as mentioned above, line broadening need not be necessarily associated with crystallite size effects alone as evidenced from the following study. Kamath and coworkers [14] have used PXRD to unravel this fact along with simulation of diffraction patterns (through DiFFAx: a program freely available for download from the web) for $MgAlCO_3$-HT.

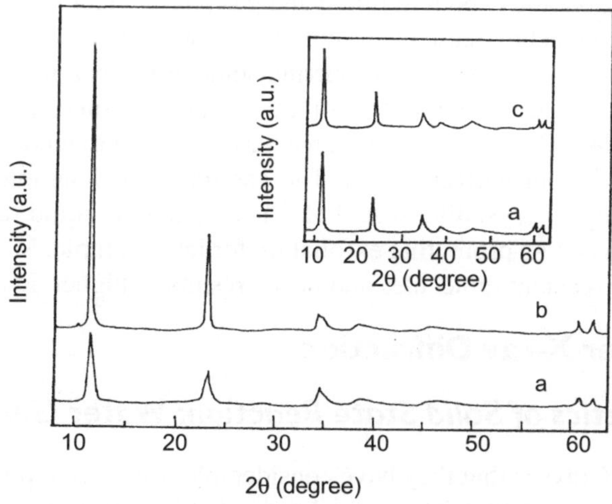

Fig. 8.13 Observed PXRD patterns of MgAl-hydrotalcite (a) aged at 90°C and (b) hydrothermally treated at 180°C (Inset is comparison between aged sample (a) and the simulated PXRD (c) pattern) (reproduced from ref. 14)

Fig. 8.13 shows the PXRD pattern of $MgAlCO_3$-HT synthesized by coprecipitation. The position of the peaks are in resemblance to that of hydrotalcite (JCPDS: 41-1428), and could be indexed with hexagonal cell, however, the reflections are considerably broadened. To verify such broadening is due to crystallite size effects, freshly prepared HT slurries were hydrothermally treated at 180°C (Fig.8.13.b). In general, it is known that hydrothermal treatment facilitates crystal growth. If one were to compare the two diffraction profiles, it could be seen that the peaks due to basal reflections (namely ($00l$)) become considerably narrower while the middle 2θ range reflections (30-50°) are not significantly influenced, which corroborates that broadening of reflections is for the reasons other than crystallite-size effects. It is known that two different polytypes in hydrotalcite occur namely $3R_1$ and $3R_2$ depending on the stacking sequence of the layers. Simulation evidenced that $3R_1$ exhibits sharp 011 reflections while $3R_2$ exhibit sharp 101 reflections in the mid 2θ region. Although the observed experimental pattern could be indexed with $3R_1$, major difference in the line shape could be seen especially for 011 reflections which are excessively broadened. Such broadening of 011 has been earlier assigned for stacking faults. Based on these observations, one could imagine and propose an intergrowth of two polytypes of hydrotalcite (thereby creating a stacking disorder). It is indeed realized by simulation wherein 40% contribution of $3R_2$ motifs within $3R_1$ polytype results in better visual match with the experimental pattern (Fig. 8.13 inset). It should however be remembered that stacking faults are not only the contributing factors for line broadening, which also occurs through various other factors such as cation vacancies, turbostraticity, and interstratification.

A detailed account of contribution of each of them in diffraction profile and line broadening has been meticulously articulated by Bookin and Drits [15]. Such disordered hydrotalcites offer better accessibility (by offering larger surface) for the reactant molecules in enabling the reaction. Tichit and coworkers [16] have recently synthesized activated hydrotalcites with such structural disorder through rehydration methodology (both in liquid and gas phase under stirring and ultrasonication). They have found a phenomenal increase in the surface area when rehydration was done under liquid medium (around 200-400 m^2/g depending on stirring/sonication) than in gas medium (15 m^2/g). A linear relationship with the activity was noted for aldol condensation of citral and acetone with the surface area of the samples. The higher surface area was attributed to tendency for exfoliation upon rehydration as evidenced from PXRD results. The gas rehydrated sample showed a sharper d_{001} reflection suggesting a higher crystallite size (141 Å) compared to liquid rehydrated under stirring (71 Å) or ultrasonication (49 Å) promoting exfoliation for latter sample. Such an exfoliation results in better accessibility of reactant molecules and hence results in higher activity.

8.2.7 In situ Powder X-ray Diffraction

Case Study 8: Kinetics of Solid State Reaction: Water Gas Shift Reaction

A unique advantage of X-rays is that they have considerable penetrating power, such that XRD can be used to study the catalysts under *in situ* conditions. *In situ* experiments were initially done by custom fabricated reactor assembly with some technical limitations. Now-a-days thanks to the advanced development in electronics and instrumentation, *in situ* reaction assembly from vendors such as Anton Paar, Austria can be easily fitted on varied diffractometers wherein any reaction gas can be allowed on the sample at temperatures even up to 1600°C. Jung and Thomson [17] were one of the first in the early 90s to fabricate such custom-designed reactor assembly to monitor the reaction through XRD under *in situ* conditions. In one of their classic study on kinetics of solid state reactions, they have followed the reduction of alumina-supported iron oxide at 675K as a function of time. Fig. 8.14-A shows the PXRD patterns of Fe_2O_3 supported on alumina under reducing conditions. At time zero or near zero, the spectrum showed reflections of α-Fe_2O_3 and Fe_3O_4. During reduction, Fe_2O_3 is rapidly converted into Fe_3O_4, which in turn is converted to Fe at slower rate. This could be evidenced by comparing the kinetics of the reaction between supported and unsupported Fe_2O_3 (Fig. 8.14-B). Unsupported Fe_2O_3 completely reduced to Fe in less than 2 min. while the supported Fe_2O_3 did not reduce to Fe even after 2h under similar conditions, evidencing the retardation of reduction by the presence of refractory support. Retardation of reduction rate upon supporting could be reasoned as follows:

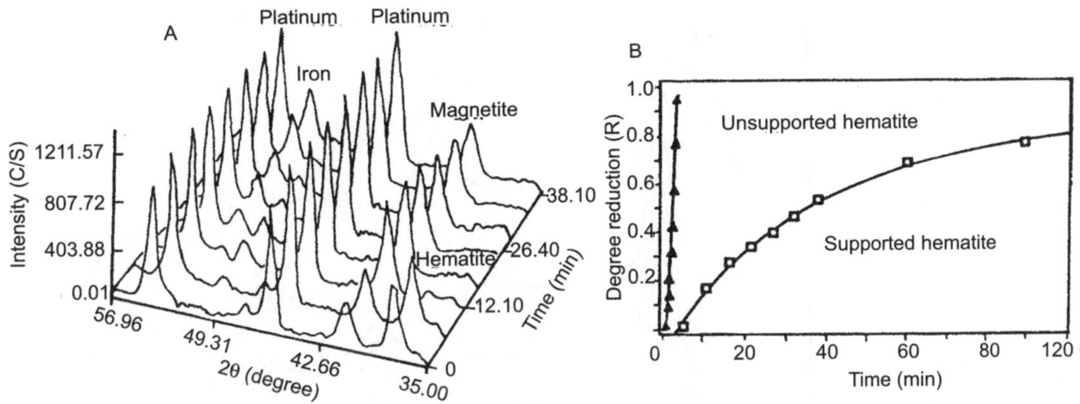

Fig. 8.14 Dynamic XRD reduction data for supported iron oxide (T = 673K, P_{H2} = 100 kPa) and (B) Hematite reduction rates (reproduced from ref. 17)

Reduction of metal oxides proceeds according to the following equation:

$$M_nO_m + m\,H_2 \Leftrightarrow n\,M + m\,H_2O$$

As the equilibrium constant for the reduction of iron oxide is of the order of 0.1, traces of water are already sufficient to oxidize the thus formed supported iron catalyst. Support plays a crucial role wherein support retains water in the form of surface hydroxyl groups, thereby effectively slowing down the reaction and preventing complete reduction. Such a study on the kinetics of reaction allows one to design appropriate reaction conditions (temperature, gas concentration, pressure) to improve the efficacy for a desired catalytic reaction. Thus PXRD study under *in situ* conditions plays an important role in the design of active catalysts and selecting appropriate reaction conditions. Such investigations will also give clues on to what extent the catalyst is stable through *in situ* phase evolution studies as presented below in another case study.

Phase Evolution: Test of Nature and Stability of Catalysts

Case Study 9: Supported Palladium Catalysts for Dehydrogenation

The remarkable activities of palladium especially on supports for hydrogenation and dehydrogenation reactions are well known in literature. It is also known that it is sometimes inadequate to characterize the catalyst at ambient conditions with the working catalyst at relatively higher temperatures and higher pressures and varied nature of atmosphere present during the reaction. At times may end up in the formation of new phases which could not be predicted or hypothesized unless one does analysis under working conditions. A classic example is use of palladium on carbon for hydrogenation of benzene [18]. The reaction was allowed and monitored through PXRD and Fig. 8.15 shows the patterns evolved during the reaction. It is quite clear from the figure, that the primarily reflections of palladium namely (111) and (200) appearing around 40 and 46.5° (2θ) decreases while the new peaks around 39 and 45.7° appear with the progress of reaction. A peak-search analysis revealed that these newly formed peaks are identified as due to β-Pd hydride (JCPDS: 18-0951C). Variation in the activity with time could well be correlated with the nature of phases evolved and PXRD played a significant role in elucidating such a phenomenon.

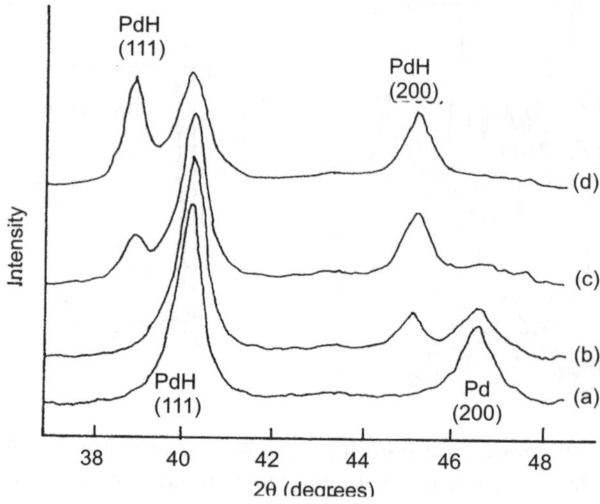

Fig. 8.15 X-ray diffraction profiles of (111) and (200) palladium lines (a) before and also in the course of benzene hydrogenation (b) 20 min; (c) 40 min; (d) 60 min.(reproduced from ref. 18)

Case Study 10: Layered Double Hydroxide for Selective Oxidation Catalysis

Layered double hydroxides are materials which have been used as precursors for the synthesis of multicomponent catalysts upon calcination resulted in mixed metal oxides with unusual property such as high surface area, homogeneity and higher dispersion and hence used for important catalytic reactions such as steam reforming, Fischer-Tropsch reactions and in environmental catalysis [19]. The important point need to be taken into consideration for such reactions over these materials as precursors is the activation conditions, like temperature, more so on the nature of atmosphere during heating. In other words one can ascertain, up to what temperature one can calcine the sample and to what temperatures it should not be. The first temperature reveals the stability of the material while the second one indicates crystallization of mixed metal oxide phases. Fig. 8.16 (A and B) shows "*in situ*" phase evolution study of CuMgAl-HT and CuCoAl-HT with increasing temperature [20]. Both the samples at room temperature showed patterns similar to hydrotalcite (JCPDS: 41-1428) showing sharp basal reflections at lower angles and broad asymmetric reflections at higher angles. With slight increase in temperature, i.e., up to 100°C, the basal reflections shift to higher angles with decreased intensity. This is attributed to removal of interlayer water molecules, thereby resulting in decreased long range ordering. A closer inspection of (003) reflection between these two samples showed a complete absence at 150°C for CuCoAl-HT while 250°C for CuNiAl-HT suggesting a higher stability of the layered network for nickel containing sample compared to cobalt containing sample. These results are in accordance with the results obtained from the thermal studies. In the temperature range 200-500°C, showed broad reflections suggesting amorphous mixed metal oxide phases irrespective of the metal ions. However, above 550°C reflections of CuO are observed. The crystallinity of this phase increased with an increase in temperature. This is attributed to sintering of CuO particles at increasing temperatures (> 600°C) resulting in poorer surface area, not suitable for catalysis. However, along with CuO, spinel phase crystallizes and one can see in the case of cobalt, crystallization of spinel started as early as 550°C while for nickel it was found around 700°C suggesting the influence of nature of metal ion in HT-like lattice in the formation of crystalline phase. Based on these results, one can conclude that an optimum temperature around 400-550°C should be ideal for synthesis of mixed metal oxides with reasonably high surface area and homogeneity, which is generally considered as advantageous parameters for enabling redox-mediated catalytic reactions.

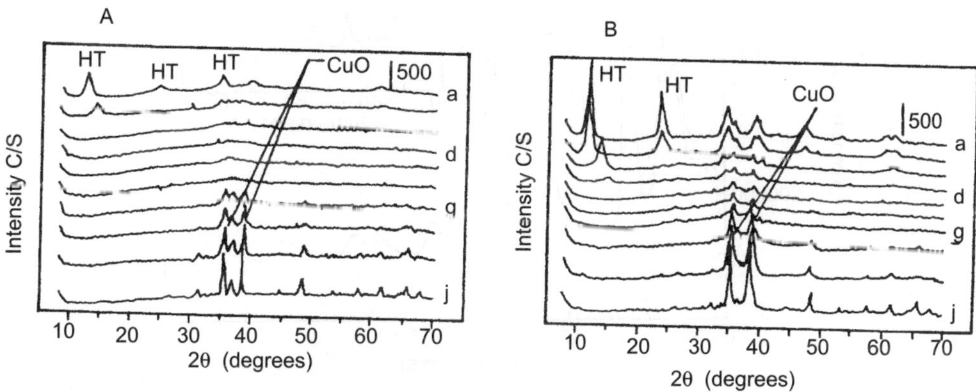

Fig. 8.16 PXRD patterns of (A) CuCoAl-HT and (B) CuNiAl-HT in situ calcined at (a) 50°C, (b) 100°C, (c) 150°C, (d) 200°C, (e) 300°C, (f) 400°C, (g) 500°C,(h) 600°C, (i) 700°C, (j) 800°C (reproduced from ref. 20)

8.2.8 XRD at Synchrotrons – Supported Catalysts with Small Metal Particles

Using synchrotron radiation as a source of XRD has the following primary advantages:
1. High intensity of the radiation gives better signal-to-noise ratio (intensity is six orders of magnitude better than the conventional tubes).
2. Collection time is significantly shorter (in seconds compared to hours in conventional tubes).
3. It is easy to vary the wavelength of X-rays (than the conventional X-ray tubes which has definite wavelength).

4. A phenomenon by which the scattering efficiency of an element decreases if the energy of the radiation is close to an absorption edge.

Such advantages can lead to identification of broader peaks from smaller particles with better accuracy. More precise *in situ* studies and distinguishing diffraction patterns of the supported particles from that of the support can also be observed. A classic example of the last point is corroborated by the case study mentioned below.

Case Study 11: Pt/Al₂O₃ Catalysts for Hydrogenation Reaction

Pt/Al_2O_3 catalysts are difficult to study by XRD, because the peaks of the Pt particles are largely obscured by those of the alumina support. However, if diffraction patterns (Fig. 8.17) are taken at two X-ray energies, one a few eV below the platinum L_{III} absorption edge at 11.564 keV and one at 50 eV lower energy, the cross section for scattering by Pt changes considerably, whereas that of support does not, and the difference between the two patterns is due to platinum particles. Fig. 8.17 inset shows the difference spectra between two signals, which indeed showed a pattern of platinum

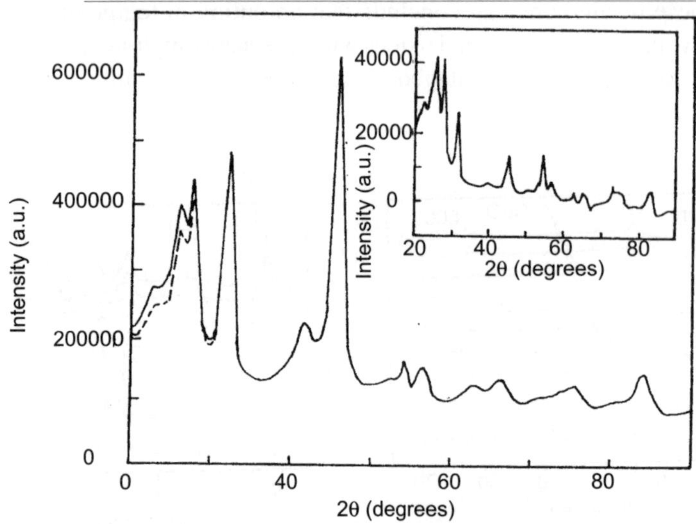

Fig. 8.17 The use of anomalous scattering in obtaining X-ray diffraction patterns of platinum catalysts supported on alumina using synchrotron X-radiation of different energies. Patterns measured well below the Pt L_{III} absorption edge at 11.508 keV (full lines) and just below the edge at 11.558 keV (dashed lines). Inset is the difference patterns of the data ($I_L - I_H$ where I_L and I_H are the measured intensities at low and high energies), where Pt particles scatter less efficiently. (reproduced from ref. 21)

particles [21]. It should be mentioned that the size of the platinum particles, (the authors have derived through hydrogen chemisorption) is as small as 1.5 nm, which would not have been possible to observe with conventional XRD. The advantages of synchrotron as a source of X-rays have been continuously growing especially in unraveling fine structures through X-ray absorption.

8.2.9 Quantitative X-ray Diffraction

In addition to the routine phase identification of materials, which is probably, one of the predominant application of X-ray diffraction technique in the field of catalysis, studies have been continuously increasing in quantifying various crystalline phases present in solids. Such studies would give appropriate clues to tailoring composition of the phases in solids and in turn on the activity for a desired catalytic reaction. In addition, similar concept is extended to assess the percentage of amorphous fraction in a crystalline material.

Quantitative analysis is based on the fact that each phase of the mixture gives its characteristic diffractogram independently of the others and the intensity depends on the amount present in the mixture. The intensity of a diffraction line (hkl) from a phase A, $(I_A(hkl))$, is given by

$$I_A(hkl) = (K_A(hkl) \times X_A) / (\rho_A) \times (\mu/\rho)_m$$

where

X_A is the weight fraction of phase A

ρ_A is the mass density of the phase A

$(\mu/\rho)_m$ is the mass absorption coefficient of the mixture

$K_A(hkl)$ is a constant for a given phase structure A, diffraction line (hkl) and experimental arrangements used

For the Bragg-Brentano geometry, which is the most commonly used geometry in most of the commercial diffractometers used today, the constant K (scaling constant) is given by

$$K = I_o \ x \ (\lambda^3/32\pi r) \times (e^2/m_e \ x \ c^2)$$

where I_o and λ are intensity and wavelength of the incident X-radiation respectively, r is the sample to detector distance and e, m_e, c are the charge, mass of electron and speed of light, respectively. A comprehensive treatment of quantitative X-ray analysis is given in literature [22]. In general two broad methodologies are employed for quantification namely:

1. Direct analysis through calibration curve obtained from the synthetic mixtures of the components under investigation
2. Internal standard addition method wherein a known weight percent of the standard material is added to the unknown sample and based on the relative intensities the concentration of the unknown is derived.

Earlier, for both analyses one or two diffraction lines were taken into consideration. However, in recent years, methods based on the use of complete diffraction pattern based on fitting experimental diffraction pattern to synthetic pattern obtained through Rietveld method have appeared. Such a method often results in improved reproducibility and accuracy. One of the representative examples is in the quantification of clay mixtures which are used as adsorbents and as catalyst supports is mentioned in the next case study:

Case Study 12: Quantification of Clay Minerals

Bish and Post studied quantification of various synthetic mixtures of clay materials such as clinoptilolite and corundum, quartz and corundum, hematite and corundum and biotite and corundum [23] with the data that were obtained using Cu K_α radiation in a diffractometer with Bragg-Brentano geometry.

Analysis of 50:50 clinoptilolite:corundum mixture illustrates the significant advantage of this Rietveld method of considering full-pattern in analyzing such complex mixtures despite the presence of severely overlapping reflections. Fig. 8.18-A is the plot of observed and calculated diffraction patterns for this mixture where more than 300 reflections were taken for quantification analysis. The analysis yielded 50.1% of clinoptilolite and 49.9% corundum, which is in excellent correspondence with the synthetic composition. Cell parameters for clinoptilolite were a = 17.6637, b = 17.962 and c = 7.4002 Å and β = 116.221° and those for corundum were a = 4.76584 and c = 12.990 Å. To appreciate this information better, an expanded region of the spectra is given in Fig. 8.18-B. The most intense doublet in this region (2θ = 21.6-22.8°) is composed of six independent reflections, a situation that would make indexing of pattern difficult even by profile refinement. Such an overlap would generally preclude highly precise determination of unit cell parameters by conventional methods, however, Rietveld analysis would do it in a more accurate manner.

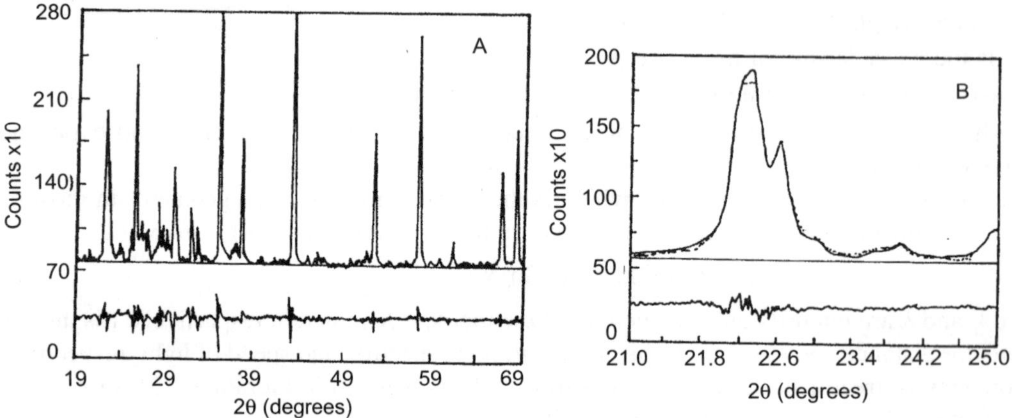

Fig. 8.18 Observed and calculated patterns for a 50:50 clinoptilolite:corundum mixture. The lower curve shows the difference between the observed and calculated patterns. Vertical marks at the bottom indicate the position of the allowed lines (reproduced from ref. 23)

8.2.10 Determination of Atomic Structure of Crystalline Catalysts

The determination of the atomic structure of a material, i.e., precise location of the position of each atom in the unit cell, generally requires a single crystal. However, heterogeneous catalysts, as the name suggests are heterogeneous polycrystalline and finely divided in nature often lack periodicity as desired for determination of complete atomic structure. However, scientists had the interest of deriving the closest possible atomic structure from powder diffraction pattern (sometimes in conjunction with neutron diffraction). This is indeed possible for materials which possess sufficient long range ordering wherein some prior information on the structure is known like in the case of zeolites. Zeolites are aluminosilicates wherein SiO_4 and AlO_4 tetrahedra are connected in three dimensional spaces (the nature of bonding results in varied nature of materials) which are highly crystalline and possess high degree of symmetry. In general, two methods are studied for structural elucidation namely:

1. Based on integrated intensities of Bragg reflections through single-crystal structure refinement methods such as the least square minimization.
2. Rietveld refinement wherein the least square refinement is performed over all data points. The entire calculated pattern is based on simultaneously refined models for the crystal structure, diffraction optics, instrumental factors and sample characteristics.

However it must be emphasized that Rietveld method is not structure solution method but a structure refinement method. Hence, a good starting method is essential for getting appropriate structural information of a solid. Veda Ramaswamy has complied steps that needed to be followed for such structure refinement nicely in a well-articulated review [24]. A classic study of structural elucidation of a solid through powder diffraction is considered in the next case study.

Case Study 13: Crystal Structure of Zeolite Omega Through Powder Diffraction

Zeolite omega, an alumino-silicate material prepared by Union Carbide, USA in 1960s, whose structure seems to be an elusive one until Alberti and coworkers have revealed their structure through high resolution powder X-ray diffraction pattern [25]. This zeolite owing to its large pore character was used in gas-oil cracking, hydrocracking and aromatic alkylation reactions. Historically two kinds of views were there on the structure. Barrer and Villiger claimed that the framework consist of gmelinite-type cages which are linked in columns parallel to the c-axis sharing their 6-rings tetrahedras. Adjacent columns, at the same height, are connected laterally form a hexagonal assemblage of columns resulting in larger channel with 12-rings and smaller channels with 8-rings. Galli published the work on natural zeolite namely mazzite whose powder diffraction pattern and unit cell parameters were very similar to omega and proposed that mazzite may be natural counterpart of structural omega. Structural information of mazzite showed subtle variations as proposed by Barrer wherein the columns of gmelinite cages are not at the same height but staggered resulting in non-planar rings. In addition, clues were not present on the location of structure directing agent (tetramethyl ammonium (TMA) ion) and Na ions in the structure. Fig. 8.19.A shows high resolution pattern of as-synthesized omega sample loaded in glass capillary and data were collected in Debye-Scherrer geometry. Rietveld refinement was performed in the space group of $P6_3/mmc$ (based on crystallographic data of mazzite).

The analysis of the results revealed that the framework of both omega and mazzite are the same despite the fact of variations in the extraframework cation content. TMA molecules are located inside the gmelinite cage, with two possible orientations which are symmetric with respect to the plane passing through the center and orthogonal to the three fold axis of the cage (Fig. 8.19.B). Sodium ions are distributed over two sites: one, at the centre of 8-ring channel hosts most of Na and is 8 fold co-ordinated to six framework oxygen and two water molecules, while the other, located along the axis of the 12-ring channel coordinates nine H_2O. In other words, one could use high resolution diffraction experiments along with mathematical treatment with some prior knowledge on the structure can refine and obtain refined structures through powder X-ray diffraction.

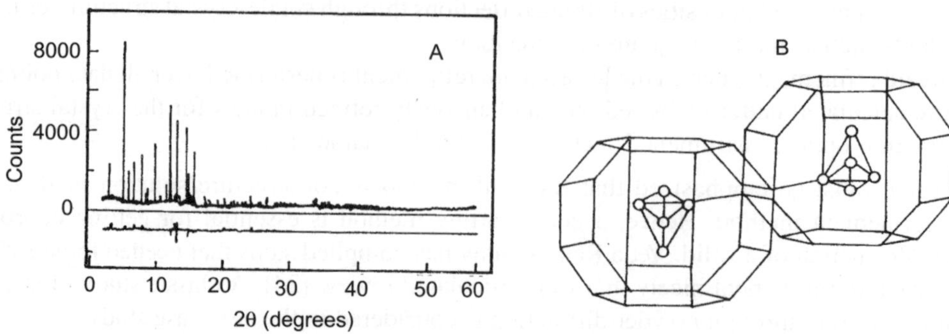

Fig. 8.19 (A) Observed (dotted upper line), calculated (solid upper line), and difference (solid lower line) powder diffraction patterns of zeolite omega. (B) Location of TMA molecules inside the gmelinite cages, shown with their two possible orientations (reproduced from ref. 25)

8.2.11 Study of Kinetics of Crystallization

X-ray diffraction is also used to determine the kinetics of a reaction such as crystallization, phase transformation etc. which will give clues on the functional and molecular aspects of the materials. Following case study will exemplify the same:

Case Study 14: Nucleation and Crystal Growth of ZSM-5

ZSM-5 is one of the well studied zeolites which have been successfully used in MTG process as well in some alkylation reactions. Derouane and coworkers [26] studied the crystallization phenomenon of ZSM-5 two decades back in a well-articulated manner which led to the discovery of microcrystalline zeolites (which are invisible for X-ray diffraction) and their unusual activity. They have studied the nucleation and crystallization of ZSM-5 using two different sources of aluminum and silicon namely Al metal and active silica and aluminum sulfate and sodium silicate and monitored the kinetics of crystallization. Fig. 8.20 shows the variation of crystallinity with time for these two sets of samples using two different templates namely tetrapropyl ammonium hydroxide (A series) and tetrapropyl ammonium bromide (B series). It may be qualitatively concluded from the figure that the nucleation and growth rates are higher for type B syntheses than for type A syntheses where a higher purity of the Al starting material seems to increase the growth rate. In addition, longer synthesis time does not affect the crystallinity of the final material irrespective of the sources employed. The variations in the rates of crystallization was attributed to difference in the interaction of TPA cations with the gel in forming zeolite crystallites wherein method B resulted in rapid formation of such crystallites, in other words rapid nucleation, however, small in nature and hence could not be detected by X-ray diffraction. However, the gel is homogeneous in nature and hence resulted in homogeneous distribution of aluminum in the crystallites and such small crystallites undergo aggregation and reordering to lead to polycrystalline aggregates.

These results are further evidenced by other physicochemical measurements such as thermal analysis and microscopy measurements. In other words, the mechanism of nucleation and crystallite growth in the formation of materials could be studied using powder X-ray diffraction which gives clues on the synthesis of gamut of microporous materials.

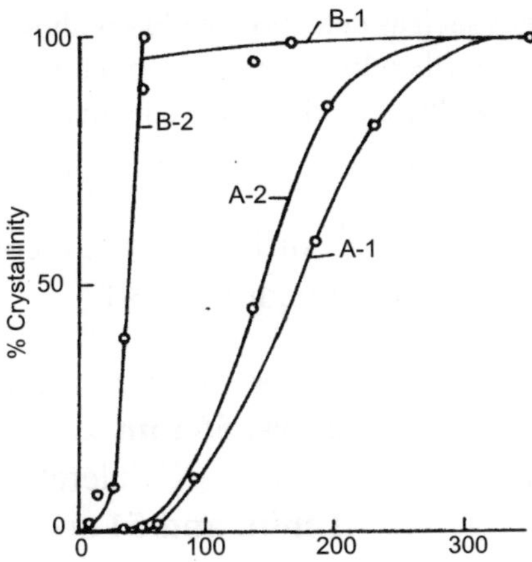

Fig. 8.20 Crystallization curves of ZSM-5 synthesis using (A) Active silica, Al metal and TPAOH and (B) Sodium silicate, aluminum sulfate and TPABr as starting materials (reproduced from ref. 26)

Based on the case studies mentioned, one can evidently perceive that X-ray diffraction is one of the essential tools for characterizing the solids, although mainly used for phase identification, specialized applications such as crystallite size measurements, atomic structure refinements, quantitative analysis, *in situ* measurements to derive information under reaction conditions, kinetics of a phase transformations, stability of materials and order-disorder phenomenon can also be studied. The instrumentation field is continuously and rapidly growing primarily due to the demands from the customers for faster and precise scanning especially those for product control applications more so under reactive conditions like in semiconductors, catalysis, pharmaceuticals. One of the very recent inclusions (in 2004) in the advancement of the instrumentation is on the detector side like X'Celerator which allows faster recording of diffractogram (a representative advantage of X'Celerator detector is shown in Fig. 8.21 where the difference between conventional detector and X'Celerator is highlighted) which allows dynamic studies in shorter timescales when coupled with reaction chamber thereby provide useful information for the design and development of catalysts. Advancements in mathematical treatment such as Rietveld method and increased computation power allows crystal structure determination of materials, which are otherwise difficult to measure, owing to their inherent characteristic of non-formation of highly-ordered or single crystals and such refinements also allow to pinpoint the location of lattice and extralattice metal ions and adsorbed/occluded molecules if any in the lattice. Efforts are also made to study the amorphous materials or less ordered materials through radial distribution function methods in conjunction with other instruments such as Extended X-ray absorption fine structure (EXAFS), neutron diffraction and small angle X-ray scattering (SAXS) where X-rays or neutrons are used as the radiation source. Use of synchrotron radiation allows us to identify small particles by alterable monochromatic X-radiation which allows the scope for the study of surfaces. Coupling of surface elucidation under reaction conditions will deliver important information which has direct bearing on catalysis science and technology. Although, diffraction methods have some disadvantages such as detection of phases above certain concentration, particles of above particular size or ordering (thereby not reflecting the real heterogeneity of the catalysts), X-ray diffraction will continuously play a predominant role in the field of catalysis.

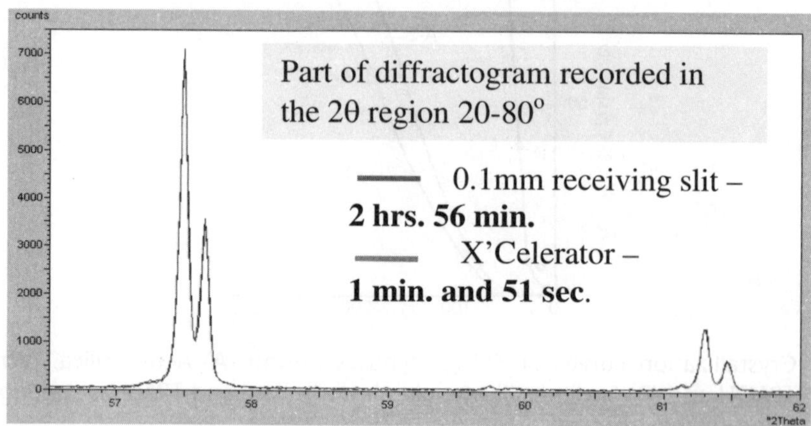

Fig. 8.21 Comparative analysis of conventional and X'Celerator detectors on a diffraction profile of a sample (reproduced from the data given by M/s Philips, India)

REFERENCES

1. W.L. van Dijk, J.W. Niemantsverdriet, A.M. van der Karan and H.S. van der Baan, Appl. Catal., 2 (1982) 273.
2. G. Faherazzi, A. Benedetti, A. Martorana, S. Giuliano, D. Duca and G. Deganello, Catal. Lett., 6 (1990) 263.
3. M.L. Ocelli, J.M. Dominiguez and H. Eckert, J. Catal., 141 (1993) 510.
4. K.J. Balkus Jr. and A.G. Gabielov, J. Incl. Phenom. Mol. Recog. Chem., 21 (1995) 159; V. Rives and M.A. Ulibarri, Coordination Chem. Rev., 181 (1999) 61.
5. S. Kannan, S.V. Awate and M.S. Agashe, Stud. Surf. Sci. Catal., 113 (1998) 927.
6. K. Bahranowski, G. Bueno, V. Cortes Corberan, F. Kooli, E.M. Serwicka, R.X. Valenzuela and K. Vcislo, Appl. Catal. A., 185 (1999) 65.
7. B.M. Choudary, N.S. Chowdari, M.L. Kantam and K.V. Raghavan, *J. Amer. Chem. Soc.,* 123 (2001) 9220.
8. D.E. De Vos, B.F. Sels and P.A. Jacobs, Advances in Catalysis, 46 (2001) 1.
9. B. Rakshe, V. Ramaswamy and A.V. Ramaswamy, *J. Catal.,* 163 (1996) 501.
10. R. Ran, X. Wu, C. Quan and D. Weng, Solid State Ionics, 176 (2005) 965.
11. F. Trifiro and A. Vaccari, in: J.L. Atwood, J.E.D. Dasvies, D.D. McNicol, F. Vogtle, J-M Lehn, G. Alberti, T. Bein (Eds.), "Comprehensive Supramolecular Chemistry, Solid State Supramolecular Chemistry: Two and Three-dimensional Inorganic Networks", Vol. 7, Pergamon, Oxford, 1996, p. 251.
12. S. Kannan, A. Dubey and H. Knozinger, J. Catal., 231 (2005) 381.
13. D. Kishore and S. Kannan, Appl. Catal. A., 270 (2004) 227.
14. G.S. Thomas, M. Rajamathi and P. Vishnu Kamath, Clays Clay Miner., 52 (2004) 693.

15. V.A. Drits and A.S. Bookin, in: V. Rives (Ed.), "Layered Double Hydroxides: Present and Future", Nova, USA, 2001, p. 39.

16. S. Abello, F. Medina, D. Tichit, J. Perez-Ramýrez, Y. Cesteros, P. Salagrea and J.E. Sueiras, Chem. Commun., (2005) 1453.

17. H. Jung and W.J. Thomson, J. Catal., 128 (1991) 218.

18. A. Benedetti, G. Cocco, S. Enzo, F. Pinna and L. Schiffini, J. Chim, Phys., 78 (1981) 875.

19. D. Tichit and B. Coq, Cattech, 7 (2003) 206.

20. S. Kannan, V. Rives and H. Knozinger, Journal of Solid State Chem., 177 (2004) 319.

21. P. Georgopoulos and J.B. Cohen, J. Catal., 92 (1985) 211.

22. R.L. Snyder and D.L. Bish, in: D.L. Bish and J.E. Post (Eds.), "Modern Powder Diffraction - Reviews in Mineralogy", Mineralogical Society of America, USA, Vol. 20, 1989, p. 101.

23. D.L. Bish and J.E. Post, Geol. Soc. Amer. Abstracts Program, 29 (1988) A223.

24. V. Ramaswamy, in: B.Viswanathan, S. Sivasanker and A.V. Ramaswamy (Eds.), "Catalysis – Principles and Applications", Narosa Publishing House, New Delhi, 2002, p. 92.

25. A. Martucci, A. Alberti, M.L. Guzman-Castillo, F. Di Renzo and F. Fajula, Microporous and Mesoporous Mater., 63(2003)33.

26. E.G. Derouane, S. Detrewriea, Z. Gabelica and N. Blomb, Appl. Catal., 1(1981)201.

Text Books on X-ray Diffraction

1. **Cullity, B.D. and Stock, S.R., 2001, Elements of X-Ray Diffraction, Third Edition, Addison-Wesley, 664 p.**
   One of the best books on X-ray diffraction with a comprehensive treatment on fundamentals, techniques, instrumentation and applications. A "must read" for those who are interested to learn or work with X-ray diffraction. *(Current retail price: $110)*

2. **Klug, Harold P. and Alexander, Leroy E., 1977, X-Ray Diffraction Procedures for Polycrystalline and Amorphous Materials, Second Edition, John Wiley, 966 p.**
   This book, which has not been revised for more than 25 years, is still the most comprehensive single-volume work on X-ray diffraction, and highly recommended for anyone who will be doing a lot of XRD work or running a laboratory. Unfortunately, the high cost of this book puts it out of the range of most students. But an excellent and "must have book" for library. *(Current retail price: $325)*

3. **Bish, D.L. and Post, J.E., Eds., 1989, Modern Powder Diffraction, Min. Soc. America Reviews in Mineralogy Vol. 20, 369 p.**
   This is a surprisingly comprehensive yet very readable volume summarizing powder diffraction feature articles from experts in the field. The first four articles alone (on principles or XRD, instrumentation, experimental procedures, and sample preparation) are worth the cost of the volume, and there is a lot more. Highly recommended and affordable. *(Available only from The Mineralogical Society of America. Price: $28)*

4. **Buhrke, Victor E., Jenkins, Ron, and Smith, Deane K., eds., 1997, A Practical Guide for the Preparation of Specimens for X-Ray Fluorescence and X-Ray Diffraction Analysis, John Wiley, 333 p.**
   Comprehensive volume (bit expensive) on sample preparation methods with discussions of sources of errors in analyses of prepared specimens for XRD and XRF. *(Current retail price: $115)*

5. **Jenkins, Ron and Snyder, Robert L., 1996, Introduction to X-ray Powder Diffractometry, John Wiley, 403 p.**

   A good introduction to XRD for new users includes good sections on instrumentation, equipment alignment, specimen preparation, and modern computer-based analytical methods; much of the training at the ICDD XRD courses is based on this volume. *(Current retail price: $105)*

6. **Pecharsky, Vitalij, Zavalij, Peter., 2005, Fundamentals of Powder Diffraction and Structural Characterization of Materials, Springer, With CD-ROM., 713 p.**

   A very recent book which provides an in-depth introduction to the theories and applications of the powder diffraction method for structure determination. This includes treatment involving atomic structure elucidation, Rietveld methods of structural deduction and refinement and advancements in instrumentation supplemented by a CD which has raw diffraction data of many materials for better understanding. *(Current retail price: $85)*

A resource page on X-ray diffraction

Matter Online Tutorial on Diffraction (http://www.matter.org.uk/diffraction/)

Excellent interactive tutorial on all aspects of diffraction phenomena, with a section concerned on XRD. Produced by a non-profit consortium of Materials Science departments in Universities in the UK. Highly recommended.

**X-Ray Diffraction Online Tutorial
(http://www.uni-wuerzburg.de/mineralogie/crystal/teaching/basic.html)**

Very mathematical tutorial from The University of Würtzberg in Germany on the theoretical basis of crystallography and crystal physics in the context of X-ray Diffraction processes. Good illustrations show how structural variations and specimen differences are manifested in the resultant data.

Scintag XRD Basics (http://epswww.unm.edu/xrd/xrdbasics.pdf)

An introductory tutorial from Scintag (the producers of diffractometer) is a good basic introduction to Bragg diffraction and powder diffractometry, and some of the advanced capabilities of XRD as an analytical technique.

**A brief primer on X-ray diffraction
(http://pubs.usgs.gov/info-handout/diffraction/html/index.html)**

A simple but quite comprehensive write-up on x-ray diffraction covering basics, techniques and applications with a focus mainly towards mineralogy, created by F.T.Dulong and J.C.Jackson at U.S. Geological Survey.

**Laboratory Manual for X-ray diffraction
(http://pubs.usgs.gov/of/of01-041/htmldocs/intro.htm)**

U. S. Geological Survey Open-File Report 01-041 is an introduction to X-ray diffraction primarily concerned with the analysis of clay minerals. Contains an intro to XRD and very extensive specimen preparation procedures for laboratory analysis.

Bragg's Law and Diffraction (http://www.bmsc.washington.edu/people/merritt/bc530/bragg/)

This introduction to Braggs Law Diffraction features a Java applet which simulates coherent and incoherent scattering. The user specifies the incident wavelength (λ), d-spacing (distance), and incident angle (θ) and the simulator shows the coherence (or lack of coherence) of the diffracted beam as a detector intensity.

Interactive Course on Symmetry and Analysis of Crystal Structure by Diffraction (http://www-sphys.unil.ch/x-ray/)

Brief but fairly thorough introduction to the diffraction of X-rays by crystalline materials. Makes extensive use of web-interactive visual aids.

Crystal Symmetry Groups (http://epswww.unm.edu/xrd/symmetry.pdf)

A very good short introduction to crystal symmetry by Robert B. Von Dreele of Los Alamos National Laboratory.

IUC Teaching Pamphlets Home Page (http://www.us.iucr.org/iucr-top/comm/cteach/pamphlets.html)

The International Union of Crystallographers has put together an excellent set of online tutorials for Diffraction and Crystallography which run from very basic to very advanced.

Crystallography 101 (http://www-structure.llnl.gov/xray/101index.html)

A comprehensive and relatively advanced (and frequently very mathematical) online crystallography tutorial from Lawrence Livermore National Lab. The emphasis is on characterization of organic molecules by single crystal techniques but the information is generally applicable to all types of X-ray crystallography.

Crystallography and XRD Links
Clay Minerals Society Links Page (http://cms.lanl.gov/site_lis.html)

A very extensive set of links which is slanted towards the analysis of clay minerals but also includes lots of more general crystallography, crystal chemistry, X-ray diffraction links and geological links. There are links to some great Earth Science sites here (and Dilbert too). Maintained by Steve Chipera of Los Alamos National Lab.

Crystallography Journals Online (http://journals.iucr.org./)

Online versions of all journals published by the International Union for Crystallography. Journals of interest include Applied Crystallography, Crystal Structure Communications, and Foundations of Crystallography. Articles from the current issues may be available online, but access to more than short summaries requires an online subscription.

Crystallograpy Worldwide Index (http://journals.iucr.org./cww-top/crystal.index.html)

IUCr-maintained page provides extensive links to other crystallography resources, software, hardware, vendors and many related information.

Silicon Valley X-ray Site (http://www.xraysite.com/index.html)

An independently maintained site sponsored by a number of XRD vendors, which provides an up-to-date links, news and invited technical articles concerned with X-ray diffraction and other types of X-ray imaging.

Determination of Pore size Distribution in Solid Catalysts

Essentially there are five different types of isotherms that are obtained for the adsorption of gases on solids. The types of isotherms (Type IV and Type V) obtained for adsorption of gases on porous solids are different from those obtained for adsorption on free surface. The differences arise from the fact that the porous solid has an internal surface so that the thickness of the adsorbed layers on the walls of the pores is limited by the width of the pores resulting in type IV and V isotherms instead of type II and III isotherms. A typical isotherm IV obtained on a porous solid is shown in Fig. 9.1. These isotherms are similar to type II isotherms except that at high relative pressures the

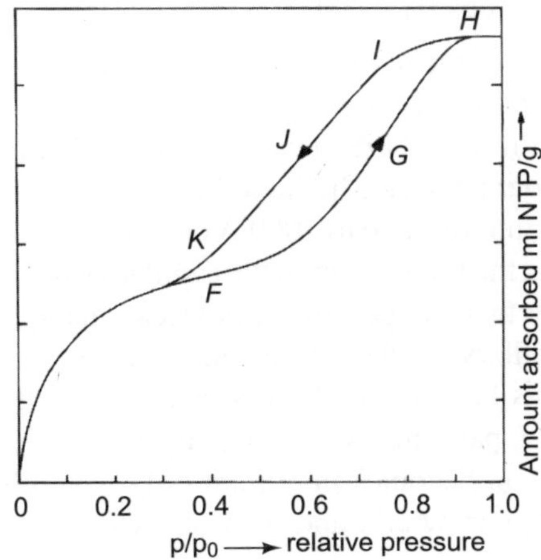

Fig. 9.1 A typical isotherm IV obtained on a porous solid

isotherm appears to reach a saturation value or increases asymptotically till it cuts the line at $p/p_o = 1$ at a small angle as compared to type II isotherms which approaches the line $p/p_o = 1$ quite asymptotically. In addition, these two isotherms often give rise to hysteresis loop of the form shown in Fig. 9.1, such that the amount adsorbed at a given relative pressure is greater on the desorption branch HIJK than on the adsorption branch FGH.

It is generally considered that type II isotherm is obtained only when the adsorbent possesses pores in the transitional range according to the definition of Dubinin. That is the pores having diameters in the range of few tens to few hundred angstrom units. The general accepted classification of porous solids is:

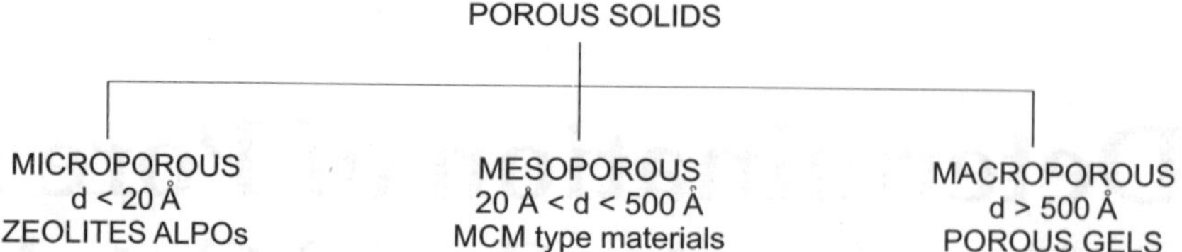

In the pores the capillary condensation accounts for the desorption branch HIJK of the isotherm. When the solid contains macropores also then the branch I'H' results.

In the low pressure region, the branch DEF is due to free adsorption as a monolayer in the walls of the pores just like the case of adsorption on free surface and is therefore completely reversible. If in addition to transitional pores, micropores are also present simultaneously, then its effect is felt in the pre-loop branch DEF itself.

The phenomenon of hysteresis in adsorption-desorption isotherms is usually rationalized in terms of capillary condensation theory. It is known that if a capillary tube is immersed in a liquid, the liquid rises in the capillary and forms a meniscus, which is more or less hemispherical and concave towards the vapour phase. The vapour pressure in equilibrium with the meniscus in the capillary is equal by an amount equal to the hydrostatic head of the column of liquid in the capillary. The vapour pressure lowering in a cylindrical capillary is given by the Kelvin's equation

$$\ln p_o - \ln P = 2 \, \sigma \, V \cos \theta \, / \, r_K \, RT \, \dots\dots\dots\dots\dots\dots\dots\dots\dots\dots\dots\dots\dots\dots\dots$$

Where σ is the surface tension of the liquid, V is the molal volume of the liquid and θ is the angle of wetting of the wall by the liquid and $= 0$ in the case of complete wetting, r_K is the radius of the capillary. Consequent to this in fine capillaries a vapour will condense into a liquid at pressures lower than the normal saturation vapour pressure. In Table 9.1 the values of P/P_o for different vapours in capillaries of different radii ranging from 10 to 1000 Å calculated using Kelvin's equation are given. This condensation of the gases in the liquid form in the capillaries is termed as capillary condensation. It was suggested by Zigmondy that in addition to physical adsorption on the walls of the larger capillaries of porous solids, capillary condensation could also occur in very fine pores such as were found by him in silica gel. This he suggested could account for the high adsorptive capacity of porous adsorbents like silica gel, particularly at low temperatures. Subsequent investigations have shown that capillary condensation becomes important only when the adsorbent has capillaries at least several molecular diameters in width, and only at pressures not very far removed from the saturation pressure.

Table 9.1 Typical values of r_k, t, r_p in Å units

p/p_0	r_k (A⁰)	t (A⁰)	r_p (A⁰)	Δt (A⁰)	$\bar{r}_k$ (A⁰)	R
0.165	10	4.74	12.5	1.21	7.1	2.318
0.332	15	5.95	17.5	1.02	11.0	2.120
0.488	20	6.97	22.5	0.89	15.1	1.980
0.927	140	16.61	142.5	0.18	125.7	1.281
0.930	145	16.79	147.5	0.175	130.5	1.274
0.932	150	16.97	152.5	0.170	135.4	1.265
0.947	190	18.25	192.5	0.140	174.2	1.214
0.949	195	18.39	197.5	0.140	179.0	1.215
0.950	200	18.52				

Typical values of r_p, t, r_k and the calculated values of R

Reprouced from Barrett, Joyner and Halenda, *J. Am. Chem. Soc.*, 73,373 (1951).

As stated earlier, the striking characteristic of capillary condensation is the appearance of hysteresis loop in adsorption – desorption isotherms as shown in Fig. 9.1. Hysteresis may be reversible or irreversible, it is reversible if on repetition of the experiment the adsorption isotherm and the loop is completely reproduced. However, it is irreversible if the second experiment gives a different curve. There has been considerable divergence of views regarding the cause of hysteresis in capillary condensation.

9.1 INCOMPLETE WETTING THEORY

Zsigmondy assumed that due to contaminations (permanent gases particularly air) adsorbed on the walls of the capillary, the condensed adsorbate does not wet the walls completely during adsorption, As the pressure is lowered, more and more impurities get displaced by the vapour and finally at saturation pressure the complete wetting takes place.

According to Zigmondy, therefore, hysteresis is not a property of the adsorbent-adsorbate system but is due to extraneous factors namely the impurities previously adsorbed on the surface. The true equilibrium curve, therefore, is the desorption branch since it represents complete wetting after the removal of the impurities. It follows that drastic evacuation and cleansing of the surface or even mere repetition of the adsorption- desorption experiment should remove hysteresis. This explanation is probably applicable for the cases of irreversible hysteresis, but obviously cannot hold where the hysteresis is reversible and permanent.

9.2 THE BOTTLE NECK THEORY

Kraemer and McBain suggested that reversible hysteresis is caused by a particular type of pore geometry – a bottle neck – pore having a narrow orifice and a relatively large body – somewhat similar to Erlenmeyer flask or like an ink-bottle. Condensation in the pores begins initially from the bottom of this bottle at a pressure given by the relation

$$P_a = P_o \exp (-2\,\sigma\,v\,\cos\theta/r_b RT)$$

where r_b is the radius at the bottom of the bottle shaped capillary. After this, the pore fills up more or less rapidly with further addition of gas, the relative pressure increasing slightly. On the desorption side, however, the pore starts emptying only when the pressure is reduced sufficiently to make the meniscus at the narrow neck unstable i.e.

$$P_d = P_o \exp(-2\,\sigma\,v\,\cos\theta/r_n RT)$$

where r_n is the radius of the neck of the pore.

Therefore one gets $P_a : P_d = \exp(1/r_b): \exp(1/r_n)$ since $r_n < r_b, P_d < P_a$.

Once the narrow neck is cleared, it unlocks the larger body of the pore and the rest of the desorption proceeds quickly. But for the bottle neck, this emptying or desorption process would proceed at a greater pressure. According to this picture funnel or V-shaped pores oriented with their large openings turned outside would not exhibit hysteresis, whereas is the inverted orientation would give rise to hysteresis.

The bottle neck theory at best provide a qualitative explanation of hysteresis. But a quantitative formulation on the basis of such a theory would be difficult. The theory relates the adsorption desorption pressures to the radii of the body and the neck of the pore and there is no reason to expect any definite relation between r_b and r_n and hence between P_a and P_d.

9.3 THE OPEN PORE THEORY

A more general explanation of reversible hysteresis has been offered by Foster assuming an open pore structure of the adsorbent. It has been shown by Foster and later by Cohan that the open pore theory leads to a quantitative relation between P_a and P_d which can be tested experimentally.

Foster suggested that hysteresis was due to a delay in the formation of the meniscus in the open pores (pores are open on both sides) during the adsorption process. At the beginning of the hysteresis region, there is an adsorbed film on the walls of the capillaries, at best only a few molecules thick. As the pressure is increased more adsorption takes place. As soon as the adsorbed film becomes thick enough to bridge the pore at the narrowest point, the liquid meniscus is formed and capillary condensation sets in immediately in that pore leaving wider pores still with only adsorbed multi-layers. At saturation however all capillaries fill and since this involves the formation of meniscus in every pore, the desorption branch obeys the Kelvin equation. Thus the ascending branch of the loop corresponds to adsorption and capillary condensation occurring simultaneously but the descending branch represents capillary condensation primarily.

Extending Foster's idea that a delay in the formation of the meniscus during the adsorption is responsible for the phenomenon of hysteresis, Cohan showed that the pressure at which the ring of multi-layers begins to form on the wall of a given capillary, that is, the pressure at which the capillary fills during adsorption, is given by the equation,

$$P_a = P_o \exp(-\sigma V/rRT) = P_o \exp(-2\sigma V/2rRT)$$

In this equation 'r' is the inner radius of the annular ring of adsorbate in the pore. If the thickness of the adsorbed layer is small compared with the radius of the capillary r_c, then 'r' can be considered equal to the radius of the capillary itself. When the capillary is full, the meniscus forms during desorption the Kelvin equation gives

$$P_d = P_o \exp(-2\sigma V/r_m RT)$$

where r_m is the radius of the meniscus. If the wetting is complete $r_m = r$.

From these equations, it follows that the radius of curvature of the liquid film corresponding to adsorption is twice as large as the desorption radius and that $\left(P_a/P_0\right)^2 = (P_d/P_0)$.

That is the relative pressure of desorption is equal to the square of the relative pressure of adsorption. If the thickness of the adsorbed film (t) is comparable with the radius of the capillary, r one must use $(r-t)$ as the net radius of the annular ring instead of r.

$$P_a = P_0 \exp(-\sigma V/(r-t)RT)$$

whereas P_d still is

$$P_d = P_0 \exp(-2\sigma V/rRT)$$

$$2\sigma V/r = [\sigma V/(r-t)] \times [2(r-t)/r] = [\sigma V/(r-t)] [(2r-2t)/r]$$

Therefore the relation between P_a and P_d becomes

$$(P_d/P_0) = (P_a/P_0)^{[(2r-2t)/r]}$$

It is easy to recognize that when $t \ll r$ then the above equation reduces to the original expression

$$(P_d/P_0) = (P_a/P_0)^2.$$

When $r = 2t$; P_d becomes equal to P_a, that is the radius of the capillary is equal to twice the thickness of the adsorbed layer, hysteresis should disappear. Since the smallest possible value of t is the diameter of a molecule, hysteresis cannot occur in capillaries narrower than four molecular diameters. It is easy to recognize that hysteresis starts in open cylindrical capillaries when r becomes greater than 2t.

One can therefore calculate the thickness of the adsorbed layer at the beginning of the capillary condensation from the relation $t = 1/2r_h$ where r_h is the radius of the capillary corresponding to the beginning of the hysteresis loop, calculated from the Kelvin equation.

The open pore theory can explain why certain adsorbates exhibit hystersis on a given adsorbent, while other adsorbates show no hysteresis on the same adsorbent. Hysteresis occurs only in capillaries more than four molecular diameter wide. It may also happen that a capillary pore is more than four molecular diameter wide for a small molecule like water, less than four molecular diameter wide for larger molecule like benzene. In such a case, hysteresis may be observed with water but not with benzene.

On this basis, it is possible to divide porous adsobents into four groups according to the average pore diameter.

1. Adsorbents having narrow pores, like chabazite and some charcoals. These substances exhibit only adsorption but no capillary condensation. Consequently, isotherms of even the smallest molecules show no hysteresis.
2. Adsorbents with somewhat wider pores, with radii in the vicinity of 10 Å. These show hysteresis for small molecules like water but none for large molecules like benzene.
3. Adsorbents with still larger pores which show hysteresis for all vapours
4. Adsorbents with very large pores in which capillary condensation occurs so close to the saturation pressure so that the hysteresis is not discernable.

It must be remarked that the open pore theory cannot satisfactorily explain the occurrence of irreversible hysteresis and certain phenomena observed by the proponents of the bottle neck theory.

There are cases where incomplete wetting may account at least part of the hysteresis. Some others where the bottle neck mechanism may be operative because of the existence of such pores cannot be ruled out. Perhaps the three theories together can give a reasonable explanation of the diverse facts of hysteresis in adsorption. Whatever be the exact cause of hysteresis a common point of all these theories is that the desorption branch of the hysteresis curve is more significant for calculating the pore radius than the adsorption branch, which in many cases has little to do with capillary condensation at least at the initial stages of adsorption.

9.4 DETERMINATION OF PORE SIZE DISTRIBUTION

The usual method of determining pore size distribution in the transitional range (i.e. 10 to 200 Å) is by the interpretation of the sigmoid shaped adsorption isotherms of nitrogen at its boiling point. However any condensable vapour can be used. The experimental method consists of first determining the adsorption branch of the isotherm upto saturation vapour pressure (i.e.$P/P_0 = 1$). At the saturation pressure the whole pore structure is filled by capillary condensation so that the volume adsorbed at this point in terns of the volume of liquid gives the pore volume. The desorption branch is now traced by lowering the pressure gradually and for any point on the desorption branch the Kelvin's equation applies. But since there can be adsorbed multi-layers on the walls of the pores before they give rise to capillary condensation, the actual pore radius is given by

$$R_c = t + 2\sigma V/RT \ln (P/P_0)$$

If t is known as a function of the relative pressure then at any value of the pressure all the pores smaller than R_c will be filled with liquefied adsorbate by capillary condensation.

The volume adsorbed at any point on the desorption branch computed as liquid volume consist of capillary condensed liquid in pores of radii $< R_c$ and V_{mt} the volume out of the multi-later thickness in pores with readius greater than R_c. That is

$$V_a = V_c + V_{mt}$$

$V_c (<R_c) = \Pi R^2 L (R) dR$ if the pores are assumed to be cylindrical. In this expression L (R) is the pore size distribution function representing the total length of the pores in the radius range R and (R + dR). Similarly one can define the surface area of the pore walls with radii greater than R_c by an equation

$$S (>R_c) = \int 2\Pi R L (R) dR$$

The corresponding total length distribution function

$$L (>R_c) = \int L (R) dR$$

For a cylindrical pore of radius R_c the multilayer thickness volume is given by

$$V_{mt} = \int \Pi [R^2 - (R-t)^2 L (R) dR = \int \Pi (2tR - t^2)L (R) dR$$

So that the volume adsorbed

$$V_a = V_c (< R_c) + tS (>R_c) - \Pi t^2 L (> R_c)$$

This fundamental equation has been used by various authors with suitable approximations so as to enable the calculation of pore size distribution by numerical integration procedure.

9.5 SHULL'S METHOD

Assuming the pore size distribution follows Gaussian or Maxwellian pattern or a combination of two i.e.

$$L\,(R) = A_M\,R\,Exp\,(-R/R_o)$$

$$L\,(R) = A_G\,exp\,-\{(B/R_o)(r-R_o)\}$$

A_M, A_G, and B are parameters peculiar to each of the assumed distribution functions. R_o is the most probable radius, A_M and A_G are dimensionless numbers determining the absolute frequency of the pore sizes. B determines the sharpness of the distribution function.

With fixed values of B and R_o one can evaluate $V_c(<R_c)$, $S\,(>R_c)$ and $L(>R_c)$ at selected values of R_c. The volume adsorbed is plotted as a function of P/P_o or R_c for each value of the chosen constants. The experimental isotherms are then matched with the theoretically computed isotherms to obtain the distribution pattern for the experimental isotherm. This method is tedious and it is a process of curve fitting and in most instances the distribution functions assumed are not sufficient to get the best fit.

9.6 BARRETT, JOYNER AND HALENDA (BJH) PROCEDURE

This procedure can be seen pictorially as shown in Fig. 9.2. Since all the pores are filled at unit relative pressure i.e. $P/P_o = 1$ by capillary condensate as well as multilayer thickness, the pore volume can be computed if one knows the capillary condensed volume by the relation

$$V_p = V_K(\,r_p^2/\,r_K^2)$$

Where r_p and r_k are radius of pole and kelvin radius respectively.

Now consider the situation when the pressure is lowered from $(P/P_o)_1 = 1$ to $(P/P_o)_2$ say $= 0.98$ the volume desorbed arises out of the evaporation of capillary condensate and the reduction in multi-layer thickness Δt_1. The pore volume of these sets of pores can then be

$$V_{p1} = R_1\,\Delta V_1\,\text{ where }R_1 = (r_{p1})^2\,/\,(\,r_{k1} + \Delta t_1)^2$$

The subscript 1 denotes the step in the desorption and also the Δ quantities refer to this step either in the reduction in the multi-layer thickness or the volume desorbed in this step due to decrease of relative pressure. When the relative pressure is further reduced to $(P/P_o)_2$, then complication arises because not only evaporation of capillary condensed liquid in the second set of pores and a reduction in the thickness of the multi-layers in the second set of pores occur, but also a second reduction of multilayer thickness in pores which were emptied in the first step also takes place.

The volume $V_{P_2} = R_2(\Delta V_2 - V_{\Delta t_2})$

Where $V_{\Delta t_2} = \pi L_1\,(r_{k1} + \Delta t_1 + \Delta t_2)^2 - \pi L_1(r_{k_1} + \Delta t_1)^2$ where L_1 is the total length of the first set of pores. $V_{\Delta t_2}$ can also be expresses as

$$V_{\Delta t_2} = \Delta t_2.A_{c1}$$

Where A_{c1} is the average area from which the reduction in multilayer thickness occurs in the first set of pores. Generalizing one can write

$$V_{\Delta tn} = \Delta t_n . \Sigma A_{cj}$$

But this equation is still unsatisfactory because A_c is not constant but varies with p/p_0. A_p the total surface area of each set of pores is constant which can be computed by the expression $2V_p/r_p$. Therefore A_p is related to A_c according to the expression

$$A_c = A_p \times (r_c/r_p)$$

or $\qquad A_p\,[(\,r_p - t_r)/r_p] = C\,A_p$

Substituting one gets $V_{pn} = R_n \Delta V_n - R_n \Delta t_n \Sigma C_j A_{Pj}$

Computations are now made by evaluating the V_{Pn} quantities using the experimental values of ΔV_n the amount desorbed and using values of R_n and Δt_n obtained from Kelvin's equation as well as suitable t curve as a function of relative pressure. The surface area of each set of pores A_p is computed by using the expression $2V_p/r_p$. This method is accurate but tediousness requires the use of a constant value of C. Secondly one has to compute the whole range of pore radius line by line and sum at each stage to get the cumulative values of V_p or A_p. A model calculation is given in Table 9.2.

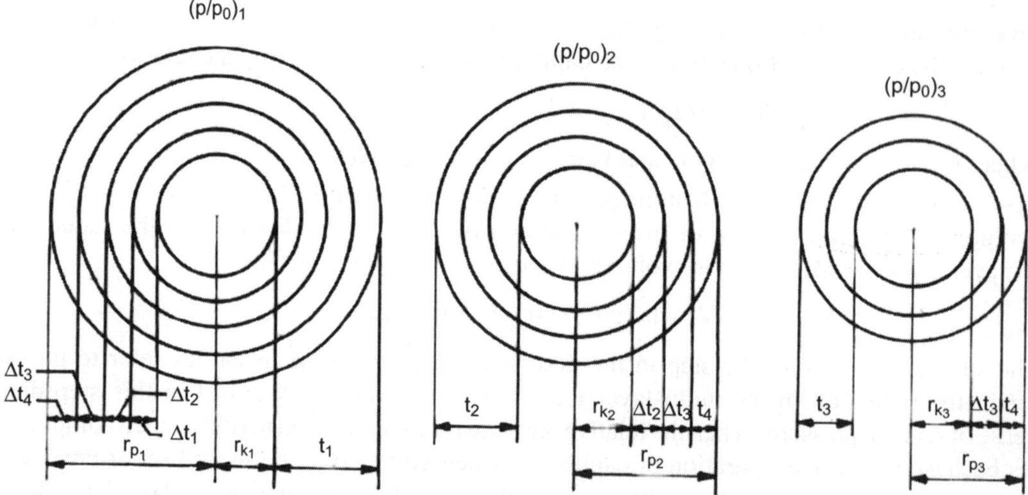

Fig. 9.2 Pictorial representation of the Barrett, Joyner and Halenda (BJH) procedure to determine the pore size distribution

Table 9.2 Computation of pore size distribution for a sample of Bone Char

p/p_0	V_p	V at STP	V_{liquid}	ΔV	r_p	$\Delta V x R$	$C^a x \Sigma AP \times 10^{-4}$	V_p	ΣV_p	A_p	ΣA_p
0.165	10	18.1	0.0282	0.0079	12.5	0.0183	0.01010	0.0014	0.2252	2.24	64.57
0.352	15	23.2	0.0361	0.0071	17.5	0.0150	0.00809	0.0040	0.2238	5.03	62.33
0.488	20	27.7	0.0432	0.063	2.5	0.0125	0.00637	0.0044	0.2194	3.91	57.30
0.580	25	31.8	0.0495	0.0061	27.5	0.0115	0.00504	0.0051	0.2150	3.71	53.39
0.642	30	35.7	0.0556	0.0055	32.5	0.0099		0.0049	0.2099	3.02	49.68
0.963	260	143.5	0.2227	0.0009	265	0.0011		0.0011	0.0029	0.13	0.26
0.964	270	144.1	0.2236	0.0008	275	0.0009		0.0009	0.0018	0.07	0.13
0.965	280	144.7	0.2244	0.0005	285	0.006		0.0006	0.0009	0.04	0.06
0.966	290	145.0	0.2249	0.0003	295	0.0003		0.0003	0.0003	0.02	0.02
0.967	300	145.2	0.2552								

* C = R X Δt X c for this computation c is taken as 0.85

Reproduced form Barrett, Joyner and Halenda, *J. Am. Chem. Soc.*, 73, 373 (1951)

9.6.1 The Method of Dollimore and Heal

The adsorption at any point of the desorption branch is shown to be

$$V_a = V_c(<R_c) + tS(>R_c) - \Pi t^2 L(>R_c)$$

$$\Delta V_a = \Delta V_c + \Delta V_m$$

$$V_m = \Delta t.S\ (>R_c) - \Pi\ 2t\ \Delta t\ L\ (>R_c)$$

Substituting $S(>R_c)$ and $L\ (<R_c)$ by summations one gets

$$\Delta V_m = \Delta t.\Sigma S(>R_c) - \Pi\ 2t\ \Delta t\ \Sigma L(>R_c)$$

Therefore,

$$\Delta V_c = (\Delta V_a)_n - \Delta t_n\ \Sigma S_P + \Pi\ 2t.\Delta t.\ \Sigma L(>R_c)$$

But we know $\Delta V_P = R_n\ \Delta V_c$ where $R_n = R_c^2/r_k^2$

Therefore, $\Delta V_{Pn} = R_n[(\Delta V_a)_n - \Delta t_n \Sigma S\ (>R_c) + 2\Pi\ t_n\ \Delta t_n \Sigma L(R_c)$

As before S and L are computed using the expression

$$S = 2\Delta V/R \quad \text{and} \quad L_p = S/2\Pi R$$

There are various other methods for the computation of pore size distribution and these procedures are only some kind of mathematical manipulation of the basic theory that has been described. The reader is referred to the original literature for the alternate methods.

9.7 THE THICKNESS OF ADSORBED LAYER IN PORES

It has been already pointed out that when capillary condensed liquid is evaporated, an adsorbed multilayer is left on the walls of the pores and the thickness of this adsorbed multilayer must be taken into account for the calculation of the pore size distribution. The true radius of the pore r_P is given by $r_P = r_K + t$ where r_K is the value of the radius calculated by Kelvin equation. This situation makes the need to know *a priori* the values of t as a function of p/p_0.

The direct way of estimating t is to measure adsorption isotherm on a non-porous reference substance which is as near as possible comparable in nature with the porous solid in question. Further it has to be assumed that at the same pressure, the thickness of multilayer evaluated from the non-porous solid is equal to the thickness on the walls of the porous solid. The value of t is obtained from the relation

$$t = (v/v_m).\ \sigma$$

Where v is the volume adsorbed and v_m is the monolayer capacity and σ is the average thickness of a single layer of adsorbed molecules. Assuming the packing to be hexagonal, and considering nitrogen as the gas that is used for the adsorption and the area occupied by a molecule of nitrogen being 16.2 Å^2 , the mass of nitrogen in grams occupying 1 cm^2 of the surface is given by

$$= (28/16.2 \times 10^{-16}) \times 6.023 \times 10^{23}$$

The liquid volume is calculated using the normal density of bulk liquid nitrogen (0.81) one gets a value of 3.5×10^{-8} cm as the value for the thickness of a monolayer i.e. the value of σ (de Boer et al reported a value of 3.54 Å while Shull has given a value of 4.3 Å on the assumption that the stacking is AAA rather than hexagonal or cubic). Since it is difficult to find a non-porous substance as comparable in nature to that of the porous system under consideration, it is usual to plot values of v/v_m against p/p_0 for a number of non-porous solids and the data points fitted into a common curve. The individual points were scattered around a composite isotherm. Wheeler, on the basis that the adsorption on the walls of fine pores is grater than on open surface at low relative pressures adopted Halsey's equation in the form

$$t = \sigma \, [5/2.303\log (p_0/p)]^{1/3}$$

and assigned a value of 4.3 Å for σ.

In fact the theory involved in each of the computation procedures for the determination of pore size distributions basically the same but the application of these methods to the same isotherm data often results in differing shapes for the distribution curves and different cumulative surface area values. These discrepancies are partly due to the different correlations used for the evaluation of the thickness of the adsorbed multilayers.

Lippens *et al* calculated multilayer thickness of nitrogen adsorbed on porous alumina using the simple equation

$$t\,(\text{Å}) = (V_L/S) \times 10^4 = 15.47 \, V_a/S$$

where V_a and V_L represent the volume adsorbed reckoned as gas and liquid respectively. S (m^2) is the BET surface area and layer thickness of 3.54 Å (5 Å) is assumed on the basis of hexagonal or cubic close packing.

Mingle and smith and Butt have stressed that the thickness of multilayers in porous adsorbents depend upon the pore structure of the solid and therefore the correlations obtained for non-porous solids were not quite applicable to porous ones. They have used the Halsey equation in the form

$$\ln\,(\,t/t_m) = (1/n)\,[\ln X\,/\ln X_m] + \ln a'$$

where n and a' are constants and X is the value of the relative pressure and X_m the value of relative pressure at monolayer saturation and t_m is the thickness of the individual layer. In the ideal case when $X = X_m$ $t = t_m$ so that ln a' should be equal to zero. However in practice surfaces deviate from ideal behaviour and therefore t/t_m will not be equal to unity when $X = X_m$ so that ln a' will have a finite value. It can therefore be considered that ln a' is a correction factor for deviation of surface from ideality.

Mingle and Smith and Butt related n and a' to the constant c of the BET equation and the mean pore radius $\bar{r}$ of the solid respectively. Mingle and Smith used a value of 4.3 Å for t_m whereas Butt has preferred the value of 3.4Å. Their equation for n and c is

$$\ln\,(1/n) = 8.38X_m - 1.40 \pm 0.03$$

Where X_m is related to c the constant of the BET equation. The relation used by them between mean pore radius $\bar{r}$ and a' is

$$\ln a' = -(3.65 \times 10^{-2})\,\ln \bar{r} + 0.11 \pm 0.01$$

A comparison of the available methods for the determination of pore size distributions is given in Table 9.3.

Table 9.3 Methods for the computation of pore size distribution in solids

Authors and essential features of the method	Basic equation proposed	Remarks
Shull Based on the assumed distribution like Gaussian or Maxwell and fitting the experimental data to these assumed distribution functions		Often the assumed distribution functions or combinations of these were not adequate. Multimode distributions could not be included in these analysis
Barrett, Joyner and Halenda A stepwise numerical integration procedure with or without approximations regarding the thinning of the multilayer adsorption on the walls of the pores.	$V_{pn} = R_n \Delta V_n - R_n \Delta t_n \sum_j c_j A_{pj}$ Approximation involving a constant value of c_j $R_n = r_p^2 / (r_k + \Delta t)^2$	The method without approximation is rather tedious
Pierce This method is also based on numerical integration as the above method with an assumed statistical number of adsorbed multi-layers	$V_p = R(\Delta V - \sum A_p \times \Delta n \times 0.23)$	This method is also tedious
Wheeler Based on his theory of adsorption on porous solids.	$\pi R_2 L(R) \Delta R = h \Delta V - h \Delta t S_g(>R) + h \Delta t\, 2\,\pi t L(>R)$ where $h\, R^2/(r-t)^2$	Numerical calculations based on this method had not been employed for the computation of pore size distribution till recently
Cranston and Inkley A more precise procedure involving tedious and lengthy calculations	$V_{12} = R_{12}(V_{12} - K_{12}) \int (r - t_{12})/2r^2)\, V_r\, dr$	This equation has been shown to be applicable to both adsorption and desorption branches of the isotherms. However the calculation procedure is time consuming.
Dollimore and Heal A step wise numerical method for the analysis of cylindrical pores. The final expression is not different from the original equation proposed by Wheeler.	$\Delta V_{pn} = R_n [\Delta V_n - \Delta t_n \sum S_p + 2\,\pi \Delta t_n \sum L_p$	This procedure tries to give a lower estimate for the total surface area contained in the pores as compared to the values obtained using BET equation.

(Contd.)

Table 9.3 Methods for the computation of pore size distribution in solids

Authors and essential features of the method	Basic equation proposed	Remarks
Anderson A straight forward method of calculation of cumulative pore size distribution curve using standard mathematical methods	$V(>R_c) = h \int R^2/(R-t)^2]G\,du$	Though an analytic method, the procedure involves evaluation of a number of integrals evaluated either graphically or numerically.
Roberts Mathematical induction method is employed to convert volume adsorbed which is expressed as a function of pore size to pore volume.	$V_{ij} = Q_{ij}(W_j - \Sigma V_j/Q_{ij})$	This method is amenable to analyze both parallel plate and cylindrical type pores
Viswanathan and Sastri An analytic method based on standard mathematical procedure and gives distributions in terms of surface area	$S(>R_c) = h/g$	This method provides the pore size distribution in terms of surface area.
Lester An abbreviated method of Cranston and Inkley.		The time involved in experimentation and the calculation sequence has been decreased to a considerable extent by this abbreviated method.
Brunauer et al. A model-less pore structure analysis method enabling the calculation of pore volume surface area and hydraulic radii		This method involves no assumptions on the model and shape of the pores and hence can be easily applied to any pore shape required.
Brunauer et al. Corrected modeless method		This procedure is mainly an extension of the previous method wherein a conversion is employed to suit the model assumed.
Brunauer et al. Micro-pore volume calculations termed as MP method		This method makes an assumption regarding the filling of micro-pores exactly in the same way as the other pores of greater radii. This assumption has been criticized.

(Contd.)

Table 9.3 Methods for the computation of pore size distribution in solids

Authors and essential features of the method	Basic equation proposed	Remarks
De Boer et al. This procedure is for the analysis of parallel plate pores	$V_{Xi} = [R\,\Delta Xi - R'_{\ xi}\sum \Delta S_{xi}]$	Based on the methods of Innes and Staggerda this method has been claimed to be accurate and precise.
De Boer et al. t-method	$t = 3.54\,Va/V_m$	Not highly suitable for the computation of pore size distribution but gives a quicker method for estimation of cumulative quantities for the pore surface area and pore volume
John and Bohra A modified form of Pierce's method involving the squares of Kelvin radii and film thickness	$\Delta V_m = U_m - F_m + f_m$ $F_m = 2V_m(0.001558)t_m\,X\,(0.23/3.6)\,x\,10^4\,R_m$ $f_m = \Delta t_m\,X\,(o.23/3.6)\sum \Delta A_i$	The pore volume obtained by this method is in good agreement with that observed experimentally. However, the surface area value obtained is lower than the value obtained by the BET method.

9.8 DETERMINATION IF MICROPORE VOLUME AND MICROPORE SIZE DISTRIBUTION

Using the original theory of Polanyi for equi-potential surfaces, Dubinin et al have derived an isotherm equation. The adsorption potential resulting form the dispersion and polar forces between the solid and adsorbate molecules are independent of temperature but varies according to the nature of the adsorbate as well as that of the solid and these forces are functions of the polarizability (α) of the adsorbed molecules.

Brunauer and his coworkers have proposed a method for the analysis of micropores (pores having widths of the order of 16 Å or less) in solids. This method is an extension of the t-method of deBoer and his coworkers and the method is named as micropore analysis method or MP method. The isotherm data obtained were converted into V_a versus t plot by replacing the p/p_0 with the corresponding t vaules. These V_a–t plots show three possibilities. (i) as long as multilayer adsorption occurs unhindered on the free surface of non-porous solid, the plot remains a straight line. (ii) if multilayer adsorption at some relative pressure begins to be augmented by capillary condensation in a porous solid, the points begin to deviate upward from the straight line.(iii) If some narrow pores are filled up by multilayer adsorption, further adsorption does not occur on the entire surface because a part of the surface has become not available. The points on the V_a–t plot then begin to deviate downward from the straight line. These downward deviations are used by Brunauer *et al* for the determination of the pore volume and pore size distributions of micropores. In Fig 9.3 the downward deviations are observed as one proceeds from t = 4 to t = 4.5Å. The straight line 2 has a smaller slope and the slope value is indicative of the surface area available for further adsorption. In other words the difference between the slope of the first and second straight line corresponds to the surface area not available for further free adsorption as a result of pore filling with multilayer adsorption.

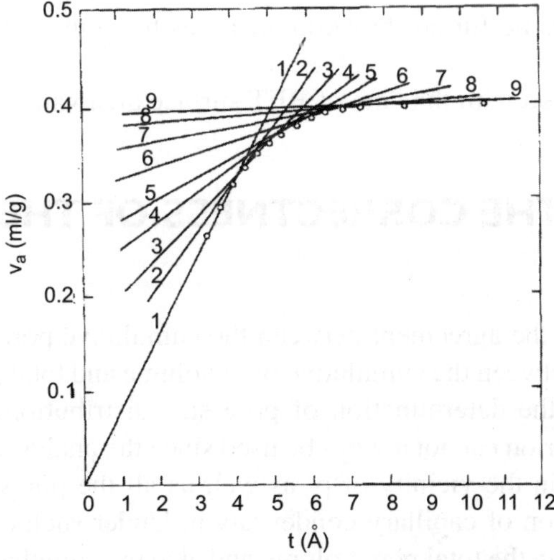

Fig. 9.3   Adsorption isotherm plot converted into a de Boer V_a – t plot. The surface areas of pore walls for the different pore groups are obtained from the defference between the slopes of straight lines 1 to 9 and the pore widths are determined from the abscissa values. This frgure n a sense illustrates the calculation of pore sizes by the MP method of micropores [From ref 14]

Table 9.4 Pore size distribution data for a sample of silica gel

Pore Group	S_{i+1} m²/g	S_i-S_{i+1} m²/g	Mean r_n	V_i ml/g
1	520	172	4.25	0.1156
2	360	160	4.75	0.0760
3	280	80	5.25	0.0420
4	200	80	5.75	0.0460
5	140	60	6.25	0.0375
6	80	60	6.75	0.0405
7	20	60	7.25	0.0435
8	10	10	7.75	0.0077
		Total 782		total 0.4088

(reproduced from R. Sh. Mikhail, S. Brunaer and E. E. Bodor, J. Colloid and Interface Sci., 26,45, 54 (1968) and also see J. Hagymassy, Jr., S. Brunauer and and R. Sh. Makhail, J. Colloid and Interface Sci., 29,485 (1969).

The volume of this group of pores is given by

$$V_1 = 10^4 (S_1 - S_2) [(t_1 + t_2)/2]$$

where S_1 and S_2 are the values of surface areas obtained from the slopes of curves 1 and 2 of V_a – t plots and t_1 and t_2 are thickness of the films in the narrowest pores of the groups. If one were to proceed from t_2 to t_3 then from the slope of the straight line 3 the surface area of the proes which are filled by multilayer adsorption in the second step is computed as $S_2 - S_3$ and the corresponding hydraulic radii is given as $(t_2 + t_3)/2$ and the corresponding pore volume is computed using the expression given above. The analysis continued until there is no further decrease in the slope of the V_a versus t plot which means no further blocking of pores by multilayer adsorption. Typical data reduction is given in Table 9.4.

Compare the data computed with the data of BET surface area value of 793 m²/g and total pore volume of 0.4034 ml/g.

9.9 CRITERIA OF THE CORRECTNESS OF THE PORE SIZE ANALYSIS

Usually the criterion used is the agreement between the cumulative pore surface and BET surface area value. The agreement between the cumulative pore volume and total pore volume is considered as the second criterion for the determination of pore size distribution. However in the case of micropore analysis, this criterion cannot always be used since the analysis leaves out the large pores i.e. those that fill or empty in the vicinity of p_0 as well as all the pores that fill or empty at low pressures i.e. below the region of capillary condensation. Under such cases, the cumulative pore volume should be smaller than the total pore volume and also the cumulative surface area should be smaller than the BET surface. Secondly, the assumptions made on the shape of the pores namely cylindrical and parallel plate pore shapes may give grossly distorted values for the volume and surface area of pores that do not have these shapes.

REFERENCES

1. E.P. Barret, L.G. Joyner and P.P. Halenda, *J. Am. Chem. Soc.,* 73,373 (1951).
2. R.W. Cranston and F.A. Inkley, Advances in Catalysis, 9,143 (1957).
3. C. Pierce, J. Phys. Chem., 57,64,149(1953).
4. G. Halsey, J. Chem. Phys., 16,931(1948).
5. C.G. Shull, *J. Am. Chem. Soc.,* 70,1405(1948).
6. D. Dollimore and G.R. Heal, J. Appl. Chem., 14,109 (1964)
7. R.B. Anderson, J. Catalysis, 3,50(1964).
8. J.B. Butt, J. Catalysis, 4, 685(1965).
9. J.O. Mingle and J.M. Smith, Chem. Eng. Sci., 16,31 (1961).
10. J.P. Irving and J.B. Butt, J. Appl. Chem., 15,139 (1965).
11. B.C. Lippens, B.G. Linsen and J.H. deBoer, J. Catalysis 3,32 (1964).
12. J.H. deBoer and B.C. Lippens, J. Catalysis, 3, 38,44 (1964).
13. J.H. deBoer, A Van der Heuvel and B.C. Lippens, J. Catalysis, 3,268 (1964).
14. R. Sh. Mikhail, S. Brunaer and E.E. Bodor, J. Colloid and Interface Sci., 26,45,54 (1968)
15. J. Hagymassy, Jr., S. Brunauer and and R. Sh. Makhail, *J. Colloid and Interface Sci.,* 29,485 (1969).
16. B. Viswanallian and MUC Saslvi, J. Catalysis, 8, 312 (1967).

Electron Microscopy

Research in the area of catalysis, in fact in general, materials, has been undergoing tremendous change due to the possibility of examining the catalytic systems with powerful microscopy which can resolve finer details and directly image the particles in the system. In simple terms, an electron microscope uses a particle beam of electrons to illuminate a specimen and create a magnified image. The resolving power of a light microscope is determined by the wavelength of the visible photons used and hence restricted to a magnification of a few thousand times. Electron microscopes can give greater resolution and magnification (nearly 2 million times) because the wavelength of the electron is smaller than that of the visible photons normally used in optical microscopes. However, it should be remembered that all types of microscopes have limitations with respect to the level of magnification. In principle, the electron microscope uses electrostatic and electromagnetic lenses in forming the image by controlling the electron beam to focus at a specific plane relative to the specimen in a manner similar to how a light microscope uses glass lenses to focus light on or through the specimen to form an image.

Historically, German Engineers, Ernst Ruska and Max Knoll built the first prototype of electron microscope in 1931 following Ruska's Ph.D thesis on magnetic lenses. However, this initial instrument was capable of magnifying objects by only four hundred times, but it demonstrated the principles of an electron microscope. Two years later, Ruska constructed an electron microscope that exceeded the resolution possible with an optical microscope. Simultaneously, Reinhold Rudenberg, Scientific director of Siemens, stimulated by family illness which necessitated making the poliomyelitis virus particle visible, patented the electron microscope in 1931. At the same time Davisson and Calbrick studied the properties of electrostatic lenses. In 1934 the possibility of surpassing the resolution of light microscope was demonstrated by Driest and Muller. In 1937 on wards, Siemens funded Ernest Ruska and Bodo Von Borries for the development of electron microscope. In 1938 Ruska and Bodo von Borries constructed the first electron microscope which showed a 10 nm resolution. About the same time, Siemens came out with a commercial Transmission Electron Microscope (TEM). It must be mentioned that about the same time, Eli Franklin Burton and others from the University of Toronto built up the first practical electron microscope. In the same decade Manfred von Ardenne pioneered the Scanning Electron Microscope (SEM). Although modern electron microscopes can magnify up to two million times, they are essentially based on the original prototype built Ruska at the Fritz Haber Institute, Berlin.

Essentially light and electron microscopes are similar in many aspects. Both of them consist of an illumination system (a light source (a lamp) or a electron source (a Filament)) that produces required radiation or electron beam and directs the beam to the specimen. Therefore, the source chamber can consist of an emitter and also condenser lens which focuses the illuminating beam (with the possibility of varying the intensity of the beam) falling on the specimen. The imaging section of the microscope consists of an objective lens which focuses the beam after it passes through the specimen and forms an intermediate image of the specimen and the projector lenses which magnify the intermediate image to form the final image which is either photographed (in conventional sense) or recorded (with digitalization).

The important differences between an optical and electron microscope are:

1. Optical microscopes employ lenses made of glass with fixed focal lengths while electron microscopes employ electro magnetic lenses whose focal length can be varied by changing the current passing through the coil.
2. In optical microscope, the magnification can be changed by switching to different power objective lenses mounted on a rotating system. It can also be done by changing the eyepieces with different power. In the TEM, the magnification (focal length) of the objective remains fixed while the focal length of the projector lens is changed to obtain the necessary magnification. Therefore, it is possible to image at varying magnifications.
3. Since the light source has a small depth field, different focal levels can be seen for the same specimen, while in TEM the large depth field ensures that the entire thin specimen is in focus simultaneously.
4. TEM has to be necessarily operated under high vacuum (due to the limitations of the mean free path of the electrons in air atmosphere).
5. There is constructional difference, in TEM the radiation source is normally at the top of the instrument while in light microscopes, the source is usually at the bottom of the instrument.

The comparison of optical and electron microscopes is shown pictorially in Fig. 10.1.

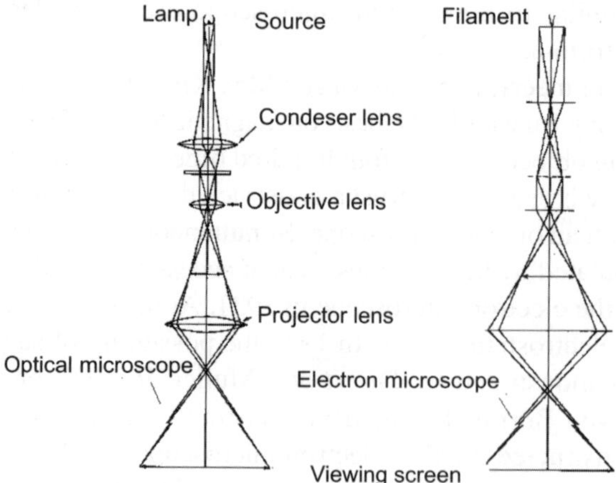

Fig. 10.1 Pictorial comparison of optical and electron microscopes

Since in electron microscopes, the electron beam is used with an accelerating potential, it is essential to know the relationship between wavelength and the potential that includes the relativity effects since the velocity of electrons can reach that of light in vacuum. The relation between λ and V (potential) taking the relativity effects into account is

$$\lambda \text{ (in nm)} = 1.23/\sqrt{(V + 10^{-6}V^2)}$$

In Table10.1, typical values of λ and the ratio of velocity of the accelerated electron to that of light in vacuum are given.

Table. 10.1 Typical values of λ and the ratio of velocity of electron to that of light in vacuum under different accelerating potential

Potential (V)	Wave length λ in nm	Velocity (10^{10}cm/sec)	Velocity of electron/velocity of light in vacuum
10,000	0.0123	0.593	0.195
25,000	0.0077	0.938	0.313
50,000	0.0055	1.326	0.414
100,000	0.0039	1.875	0.548
1,000,000	0.0012	5.930	0.941

When electrons strike a specimen various events can take place. These are shown pictorially in Fig.10.2.

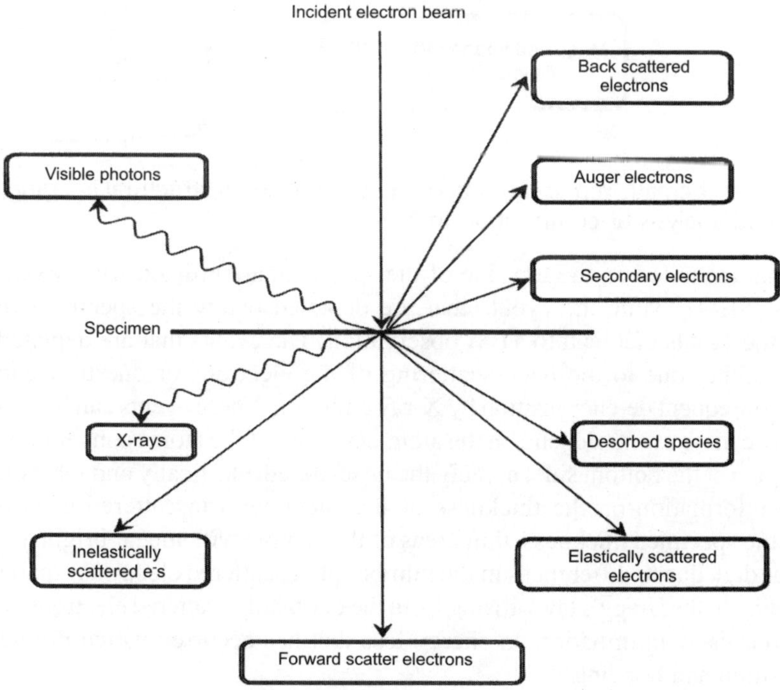

Fig. 10.2 Schematic diagram of the events that occur as a result of electron beam sample interactions

There are various ways these interactions can carry information of the sample. The type of information that can be obtained is pictorially shown in Fig.10.3

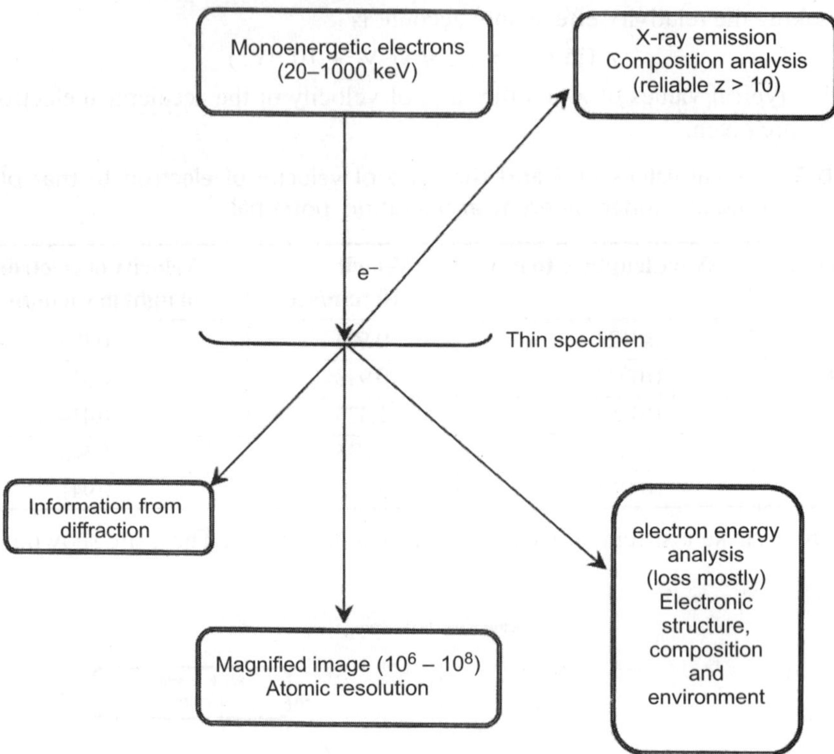

Fig. 10.3 The schematic representation of the possibilities of structural imaging, diffraction and analysis of composition in TEM

The events that are noted on the top side of the specimen are utilized when examining thick or bulk specimens (SEM), while the events that are depicted below the specimen (mostly in thin specimens) are the events that lead to TEM observation. The events that are depicted on the top of the diagram are either due to the back scattering of the electrons or due to secondary electron emission or the consequent de-energization by X-ray emission. These events can be used for chemical identification since these events depend on the atomic number of the atoms constituting the specimen. The events depicted at the bottom side, namely the unscattered, elastically and inelastically scattered electrons carry information on the thickness of the specimen (unscattered electron intensity is proportional to the specimen thickness, thin areas of the sample will appear bright and thick sample areas will appear dark due to differences in the number of unscattered electrons) spatial arrangement of atoms (according to the Bragg's law) arising from the elastically scattered electrons and inelastically scattered electrons carry information on energy loss that has occurred which provide information on both composition and bonding.

Therefore various forms of microscopy are possible depending on which of the electron beam is used for imaging. The transmitted electron beam forms an image in the Transmission Electron

Microscopy (TEM). In the operation of TEM, resolution is limited by spherical aberration but in these days aberration correctors have been able to partially overcome spherical aberration to increase the resolution. Software based corrections have also been included in the modern High Resolution TEM (HRTEM). These techniques have been successfully employed with sufficient resolution to identify carbon atoms in diamond lattice with atom separation of 0.89 angstrom. This ability to determine the positions of individual atoms within a given material has made HRTEM a valuable tool for the study of nano materials.

The scanning Electron Microscope (SEM) produces images of the specimen by the focused electron beam that is scanned across an area (usually rectangular) of the specimen. The energy loss experienced by the electron beam manifests as heat, or emission of secondary electron, light emission (cathodluminescence or X-ray emission). The display of the SEM maps the varying intensity of any of these signals into an image corresponding to the position of the beam on the specimen when the signal was generated. Hence, the image resolution in the SEM mode is about an order of magnitude poorer than that of TEM. However, since the SEM is based on the surface processes that take place rather than the transmitted beam, it is possible that bulk samples can be imaged and the 3D structure of the sample can be analyzed.

Another variation is the Reflection Electron Microscope (REM) wherein neither the transmitted nor secondary electron beam is picked up for image formation but the reflected beam of elastically scattered electrons is detected in the image. Hence this technique is usually coupled with high energy electron Diffraction (HEED).

The type of information that can be obtained from microscopic (especially TEM) examination of catalyst systems includes:

1. Morphological details like the size, shape and arrangement of the particles which make up the specimen as well as their relationship to each other on the scale of atomic diameters.
2. Crystallographic information namely the arrangement of atoms in the specimen and their degree of order, detection of atomic scale defects.
3. Possibly the compositional information especially in chosen areas and the relative composition if it is a multi-component system can be obtained from TEM observations.

Thomas and others claim that no other single instrument rivals the electron microscope in the wealth of structural (atomic, nanoscopic, mesoscopic and microscopic) topographic and electronic information that this technique provides in the characterization of solid catalysts. They have illustrated the use of these techniques in a variety of cases that include (i) pin pointing the location and topography of nano particle catalysts; (ii) constructing elemental maps and hence two dimensional compositional analysis of catalysts; (iii) *in situ* investigations of the active sites of the catalysts and the atomic level mechanistic details of the reactions taking place on the surface; (iv) elucidation of the intergrowth of similar structures (v) Chemically identifying groups of atoms of high atomic number supported on high area solids and (vi) characterizing nano particles of supported systems. All these studies can provide information on structural, electronic, or compositional details on functional catalysts, which will be useful for understanding and designing of active catalytic systems.

The details of the electron microscopic images can be clearly discernable if the sample preparation is carried out properly. There are a variety of methods that have been adopted in practice. The methodology required may vary with respect to the type of samples. It is essential that the samples are completely devoid of water (moisture) and a variety of dehydrating sequences can be followed including freeze drying and also use of volatile organic solvents. The sample either embedded or fixed has to be polished to a mirror like finish with the use of ultra-fine abrasives. The polishing has to be done carefully so that scratches and other polishing artifacts are reduced to a great extent. In order to produce thin specimens, it is possible to section slices of the material with the use of an ultra-microtome with a diamond knife. The surface can be cleaned in a variety of ways including ion etching. It is often necessary to give an ultra thin coating of electrically-conducting material, deposited either by vacuum evaporation or by sputter coating on the sample. This is done to prevent the accumulation of static electric fields at the specimen due to the irradiation with electrons required for imaging. Such coatings include gold, gold/palladium, platinum, tungsten, graphite etc. and are especially important for the study of specimens with the scanning electron microscope. Another reason for coating, even when there is more than enough conductivity, is to improve contrast.

Most of the present day catalysts are either powdered solids consisting of one or more distinct phases or supported metallic (mono or multi-metallic) systems. The supports normally employed are materials like carbon, oxides, chacolgenides or halides. The information that is normally probed in catalysis is to identify the nature of the active sites that are responsible for the promotion of the surface transformation. This can either be the nature of the chemical species involved, the crystallographic orientation or the particle shape and size of the aggregated active systems. For obtaining this type of information on catalytic systems electron microscope has become an indispensable tool especially since many of the active phases of the catalytic systems are usually in the nanoscale.

SEM can be effective and especially applicable in various areas of heterogeneous catalysis especially in the case when usual investigation methods could not provide necessary information. SEM results when appropriately combined with other spectroscopic, diffraction, adsorption and other microscopic methods can contribute to the explanation of the formation of the active structure formation, (the so called active centres) *in situ* transformation from one phase to another or identification of the simultaneous presence of multiple phases, for porous systems, details on the textural variations, and the causes for catalyst deactivation and aging. In essence, SEM observations of catalyst systems can be used to decide on the choice of catalyst material, the influence of preparation methods on the catalyst performance, activation and active phase formation and structure, texture and morphology of the catalyst systems as they are prepared, or during their use in catalysis *(in situ* studies). Some typical examples for these applications will be considered briefly.

10.1 ELECTRON MICROSCOPIC STUDIES ON CARBON AND CARBON SUPPORTED CATALYTIC SYSTEMS

Carbon has been extensively studied due to the possibility of generating all kinds of architectures starting from bulk 3D carbon materials to one dimensional nanotubes. Among the various possibilities, Black pearl, Vulcan XC 72 and CDX 975 are the typical carbons that have been employed for a variety of applications, including as catalyst support and also as electrode material for supporting noble metal catalysts. The Field emission SEM pictures of these commercial carbons are given in Fig. 10.4.

It is seen that most of these conventional carbons have spherical morphology and hence, they have been always employed for supporting active metallic components to be employed as catalysts. In contrast to these conventional carbons, carbon obtained from natural sources can assume different morphologies depending on the activation procedures adopted for generating carbon materials.

One such example is the carbon obtained from the plant source *Calotropis gigantea*. This morphology and texture of the material obtained have been examined after activation by variety of agents. Details of the surface morphology as well as the elemental composition of the activated carbon materials from *Calotropis gigantea* using carbonate and oxalate activation were obtained from Field emission gun scanning electron microscopy (FEG-SEM) studies. The SEM images of carbonate activated carbon, designated as Cg carbonate recorded at two different magnifications namely 2400 X and 5000 X are shown in Fig. 10.5.

The Cg carbonate possesses a unique morphology. The carbon materials comprises of platelet like carbon particles with evenly distributed spherical pores. Such morphology is unique to the carbon precursor, calotropis gignatea and has not been reported previously with any other lignocellulosic carbon precursor. The energy dispersive X-ray analysis indicated a carbon content of 92.62 wt.% and an oxygen content of 7.38 wt.% Unlike carbonate activation; oxalate activation resulted in distinctly different carbon morphology. Activation with $Na_2C_2O_4$ resulted in a tubular morphology. Bunches of carbon micro tubes with open ends were formed from the carbon precursor upon oxalate activation (Fig.10.6). Energy dispersive X – ray analysis showed a carbon content of 92.49 wt.% and an oxygen content of 7.51 wt.%. Thus it is evident that the carbon morphology derivable is a function of the type of activating agent. Also the carbon precursor too dictates the morphologies achievable. These studies may be expected to throw light on the reasons why carbon materials are capable of exhibiting a variety of morphologies?

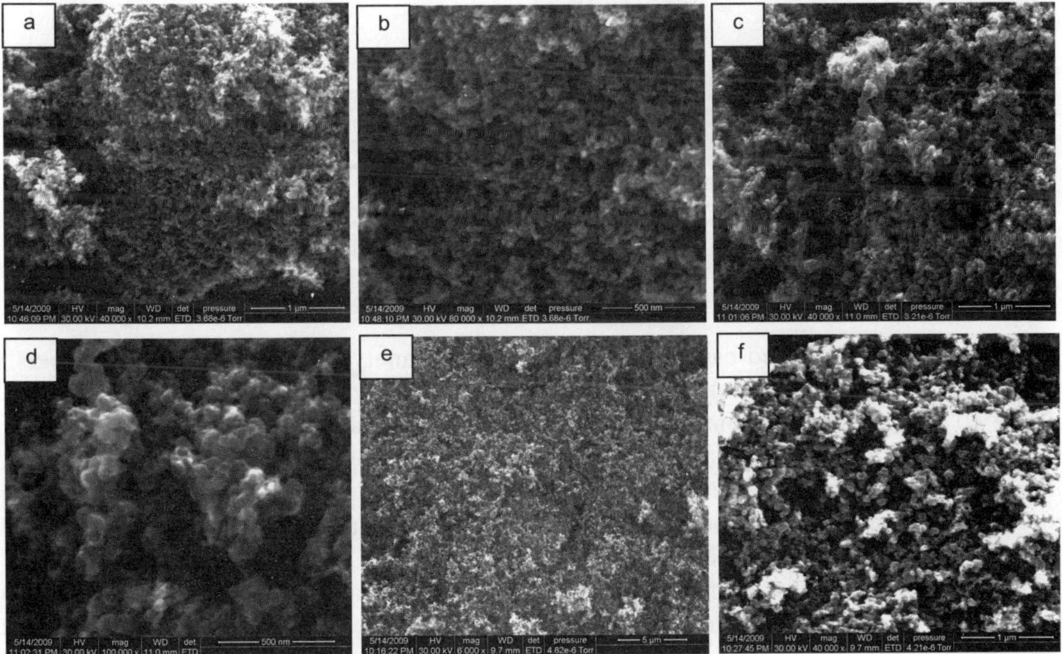

Fig. 10.4 FEG SEM images (at different magnifications) of commercial activated carbon material: (a and b) Black Pearl 2000 (40, 000 X, 80, 000 X), (c and b) Vulcan XC 72 R (40, 000 X, 1, 00, 000 X) and (e and f) CDX 975 (6000 X, 40, 000 X)

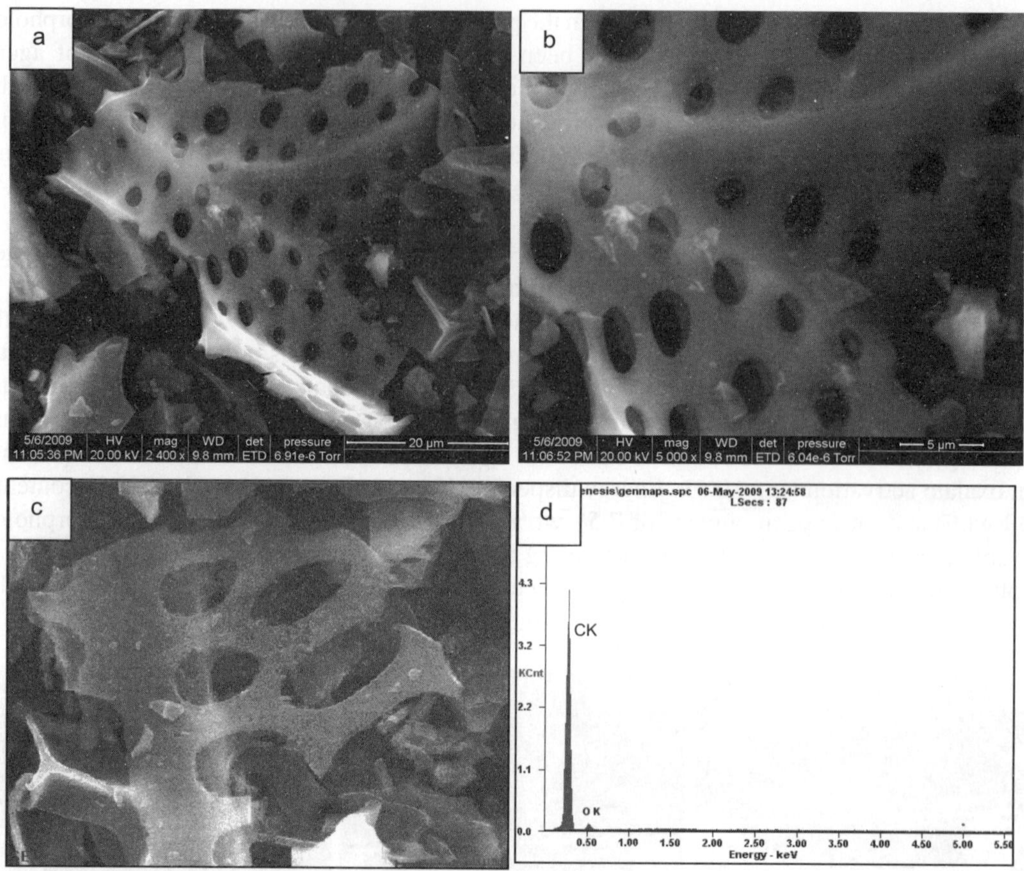

Fig. 10.5 FEG SEM images (at different magnifications) of activated (carbonate) carbon material from Calotropis gigantea, Cg carbonate, along with EDS spectrum: (a) 2400 X, (b) 5000 X and (c and d) FEG SEM image with EDS spectrum

The commercial activated carbon materials too were analyzed by EDS (energy dispersive spectroscopy) and the carbon and oxygen contents present in the carbon materials are summarized in Table 10.2. The carbon content of the plant based activated carbon materials produced through simple means is comparable to that of the commercial carbon materials. It is evident that the carbon content in any of the carbon materials may be more or less the same, however the morphology and texture could be altered and hence carbons of varying surface area and porous structure can be generated for use as supports in catalysis.

Carbon nanotubes (either single walled (SWNT) or multi walled (MWNT)) have remarkable thermal, electronic and mechanical properties arising from their structural similarity to graphene sheets and also the one dimensional character. In the nanoscale, the one dimensional character imparts some unique features which make these materials widely applicable in a variety of applications. The single walled carbon nanotube is the simplest form of nanotube and can be visualized as a single sheet of grapheme rolled-up into a steamless tube that may be several microns in length and as small as 0.7 nm in diameter. Carbon naotubes are conventionally synthesized by techniques like arc discharge, laser vapourization and catalytic pyrolysis. The carbon nanotube materials obtained employing any of these methods require post synthesis manipulation. The commercialized (by Carbolex) SWNT prepared by arc discharge using the catalyst Ni-Y (4:1 atomic ratio) has been used extensively in fundamental and applied research. The focus of the studies has been to examine the morphology and purity of the material as a result of several treatments.

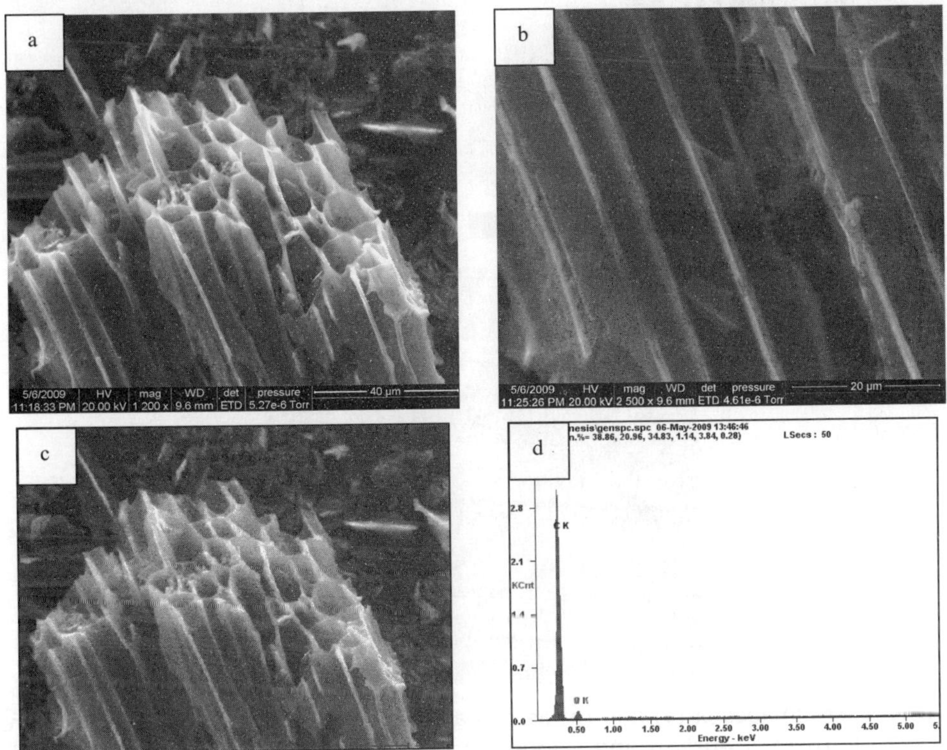

Fig. 10.6 FEG SEM images (at different magnifications) of activated (oxalate) carbon material from Calotropis gigantea, Cg oxalate 4, along with EDS spectrum: (a) 1200 X (top view), (b) 2500 X (lateral view) and (c and d) FEG SEM image with EDS spectrum

Table 10.2 Elemental analysis of activated carbon materials (EDS analysis)

S. No.	Carbon Material	Element (wt.%)	
		Carbon (C)	Oxygen (O)
1	Cg carbonate 3	92.62	7.38
2	Cg oxalate 4	92.49	7.51
3	Black Pearl 2000	93.82	6.18
4	Vulcan XC 72 R	97.7	2.3
5	CDX 975	96.88	3.12

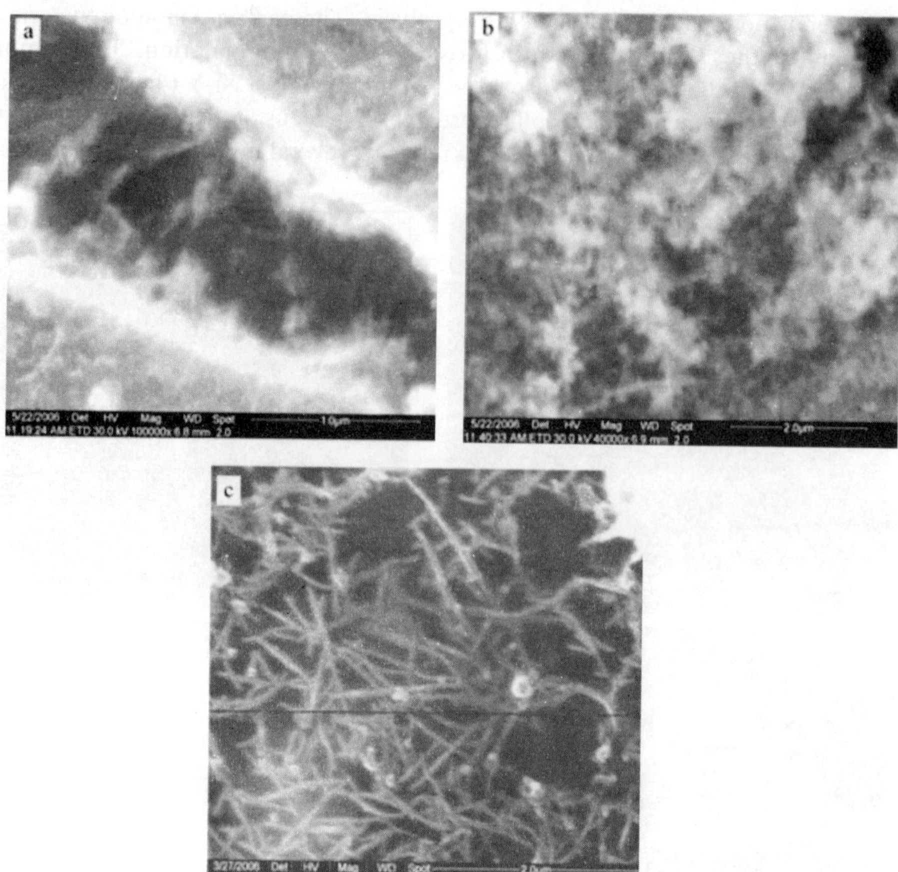

Fig. 10.7 SEM images of (a) and (b) as prepared and (c) purified SWNT (From reference J. Thermal Analysis and calorimetry, 88 (2007) 885)

The SEM images of the as prepared SWNT (in a and b in Fig. 10.7) show that the nanotubes are thin and large particles and aggregates appear to have clumped together in nanotube ropes. The purified material also still contains the particles (Fig. 10.7 (C)) but however the ropes are thicker and shorter. These changes have taken place due to the combustion process in the purification step which has combusted the amorphous carbon and the metal particles are digested during the acid treatment.

Similar or analogous changes can be expected to take place with respect to the different purification procedures that are normally adopted in the case of carbon materials. In Fig. 10.8. the SEM images of Carbon nanotube bundle and single nanotube are shown. The typical value of the diameter of the carbon nanotubes and the transparency of the nano tubes are clearly visible.

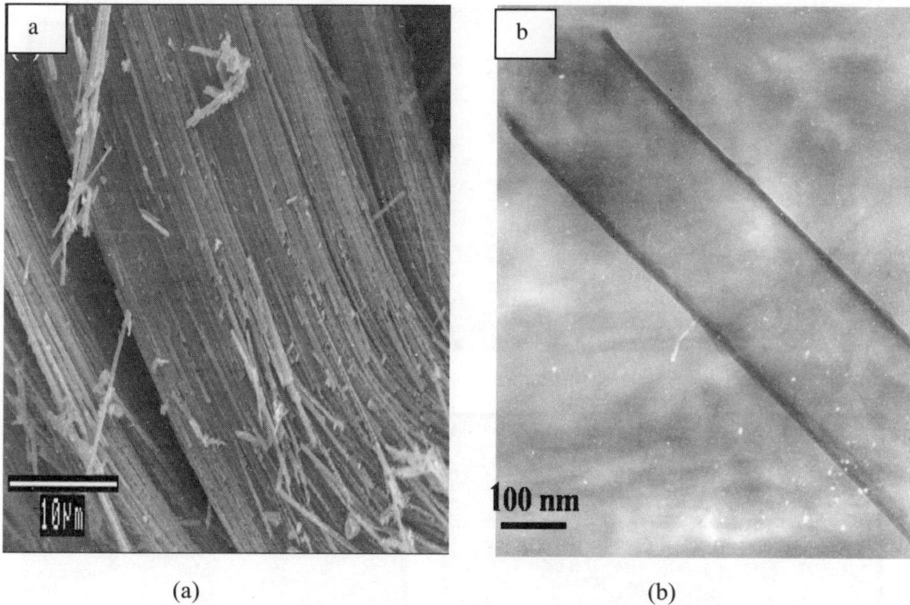

(a)   (b)

Fig. 10.8 SEM images of (a) CNT bundle and (b) single CNT showing the contours and also the dimensions

Carbon Nanotubes (CNT) and other carbon materials are conventionally employed as supports for catalysts. The metal loaded carbon nanotubes are extensively employed as electrodes for fuel cell applications. Since mostly noble (mono-, bi- and multi-) metals are employed for these applications it is essential that the dispersion of the mono metals has to be optimum and the metal particle sizes should be in nanometer range. In the case of bi- or multi-metallic systems, there should be dispersion of the metallic species such that the bi functional activity is manifested suitably. For example in the case of Pt-Ru catalyst system for the fuel cell application, the Pt species should be uniformly distributed with adjacent Ru species so that the poisoning of the Pt sites by carbon monoxide can be easily removed by oxidation of these adsorbed carbon monoxide by the OH species adsorbed on adjacent Ru sites or other catalytic sites like WO_3 which promotes the oxidation of adsorbed carbon monoxide.

The electron micrographs of metal loaded on carbon nanotubes are shown in Fig. 10.9. It is seen that the noble metals (mono- and bi-) supported on carbon nanotube systems are dispersed uniformly with respect to size and shape. The particle size of the metals dispersed on the carbon nanotubes is nearly 2-3 nm. The type direct visualization of the dispersed metals and also other systems like Pt-WO_3 systems can be directly imaged and thus maximum utilization of the noble metals can be ensured for electrode applications.

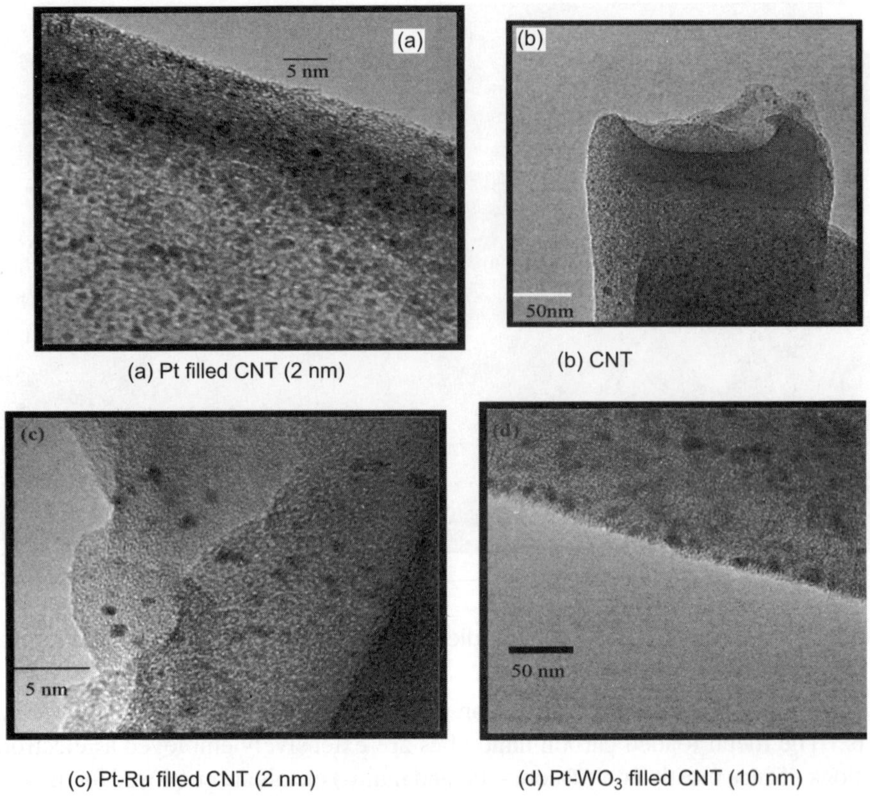

(a) Pt filled CNT (2 nm)

(b) CNT

(c) Pt-Ru filled CNT (2 nm)

(d) Pt-WO$_3$ filled CNT (10 nm)

Fig. 10.9 Noble metal dispersion on carbon nanotubes (particle sizes are indicated)

10.2 ZEOLITES AND RELATED MATERIALS

Zeolites for catalytic purposes are usually synthesized as a fine powder of approximately a few mincrometer in particle size. In addition, the clusters that are introduced in the zeolites are not distributed uniformly and hence electron microscopic examination of the zeolites is important for structural and dispersion information. Secondly, the crystals of zeolites contain defects and hence give rise to diffuse scattering or streaks in the microscopic patterns. Zeolites form a variety of families in which the common structure building units are found. The families have a strong tendency to form intergrowths between different structures within the same families. Hence it is necessary to make a two step analysis namely the structural unit as a rod or as a sheet and then to determine the relative spatial relation between the neighbouring atoms. It is important to have a clear understanding of the surface structures of zeolites so as to understand the chemical reactions which take place on the surface of zeolites and to be able to determine the elemental structural unit for the crystal growth. In spite of this situation, there are only a few studies on zeolites due to the difficulty to obtain good quality TEM images that can be processed to obtain the exact information. In the case of zeolites, it is common that one makes a comparison of the observed image with the simulated images of the ring structures. In Fig. 10.10, the TEM image of zeolite LTL along the [001] direction is shown. One can notice that in the left the simulated drawing of the idealized LTL structure is given. The HRTEM image given in the right is mostly indistinguishable from that of the simulated picture given in the left. In many of these studies, the important aspect that has been resolved is that depending on the synthesis conditions employed the zeolites contained coherent, or recurrent or random intergrowth of one or more zeolites.

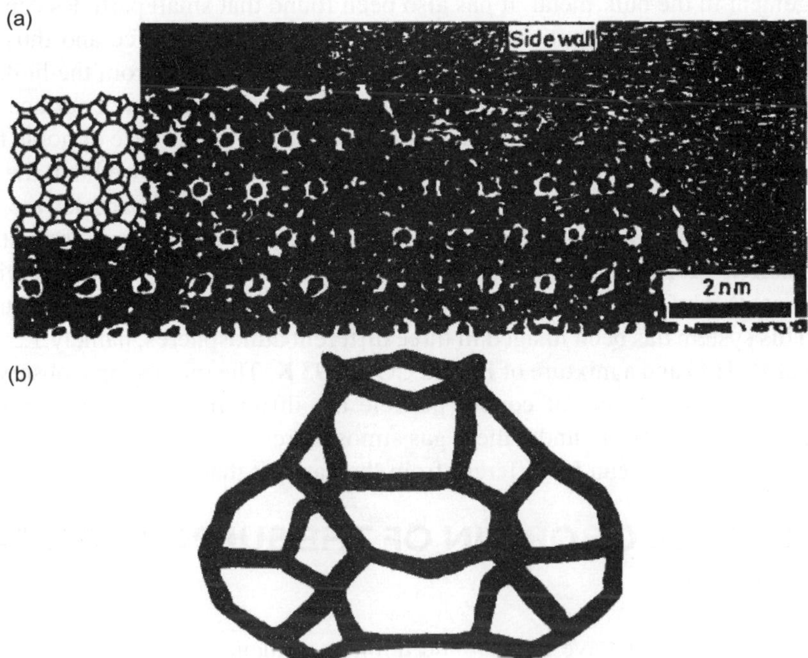

Fig. 10.10 (a) HRTFM image of Zeolite LTL along the [001] direction, (b) a schematic drawing of the framework of the idealized LTL sturcture (reproduced from J.M. Thomas and P.L. Gai, Advances in Catalysis, 48 (2004) 186)

It is also possible to sustain clusters of atoms in the zeolite channels, their position and size have also been evaluated form the microscopic examination. The position of the metal clusters in zeolite channels can be considered as the active sites for catalytic reactions and it is known that most of the zeolites contain a number of possible sites for subtending the metallic clusters. Hence the identification of the metallic clusters and their location in zeolite channels assume importance. It must also be mentioned that zeolites are sensitive to electron beams used for microscopic observations. Hence studies of mechanism of damage and damage processes have also been investigated.

10.3 SUPPORTED CATALYSTS

It is known that most of the practical catalysts are supported systems for various reasons. Hence these systems have been extensively investigated by electron microscopy. For the concept of strong metal support interaction which is normally associated with low and high temperature reduction of metals supported on oxides (typically TiO_2), *in situ* TEM observations showed that the Pt metal particles had lower contrast that can be attributed to raft like or pillbox structures. When metal particles are supported on oxide supports, the information that are often sought after are the interface, the size and the geometry of the metal particles and the shape of the small metal particles. Even though it has been always the desire to see individual metal atoms, imaging of the clusters of atoms has been often possible depending on the nature of the support. The most common particle shape for fcc metals like Pt or Au is cubooctahedral exposing (111) or (100) surface facets, following the atomic arrangement in the bulk metal. It has also been found that small particles can adopt a range of internal structures with different arrangements such as bcc, hcp or fcc and thus the reactivity and nature of active sites of the supported metal systems are different from the bulk metal.

Morphological details of catalytic significance include the distribution of surface steps, particle faceting and the nature and geometry of surface reconstructions. In the case of noble metals like Rh on silica surfaces, oxidation followed by low temperature reduction, showed that the surface of the metal particles become rough, the nature of surface sites in these metal particles are not a function of the size alone, in similarly sized particles one can have different surface coordination leading to differences in activity. It has become possible to *in situ* image the catalyst under varying conditions and atmospheres. A Cu/ZnO methanol synthesis catalyst has been imaged as a function of gas atmosphere. This system has been imaged in three different atmospheres, namely 1.5 mbar of H_2, in a 3:1 mixture of H_2:H_2O and a mixture of H_2 and CO at 493 K. The microscopic observations clearly showed that not only the facets of copper particle are different but the size and shape of the particles are also totally different under these gas atmospheres.   This shows that the active catalysts and catalytically active sites can be different from the catalyst that has been designed.

10.4 INFLUENCE OF ORIGIN OF THE SUPPORT ON CATALYST ACTIVITY

The origin of the support can have explicit and implicit influence on active centre formation. The Fe/MgO catalyst for the synthesis of higher hydrocarbons by hydro condensation of CO shows different activity depending on MgO source. Fe/MgO catalyst with MgO obtained form magnesium hydroxide has poorly ordered structure whereas the Fe/MgO catalyst with MgO obtained from a basic magnesium hydroxycarbonate has plate like structure. The Fe/MgO catalyst with MgO obtained from magnesium oxalate has clearly formed cubic structure. The catalyst with well ordered crystal structure, as in the catalyst derived from magnesium oxalate has a good selectivity toward lower olefins. The corresponding micrographs that have been used to deduce these aspects are shown in Fig. 10.11.

10.5 DEACTIVATION OF CATALYSTS

There are various methods by which a catalyst can be deactivated. The Pt-Rh catalyst used for ammonia oxidation in relation to manufacture of nitric acid showed that the deactivation is due to the accumulation of foreign metals like Fe, Ni, Ca and Mg on the Pt-Rh catalyst on gauze. In the case of hydrodesulphurization cobalt Moly catalyst, the preferential evaporation and condensation of MoO_3 has been identified to be the cause of deactivation. In the case of Ni/Al_2O_3 catalyst the deactivation is identified to be due to the irreversible binding of the active component Ni with alumina to form Nickel aluminate.

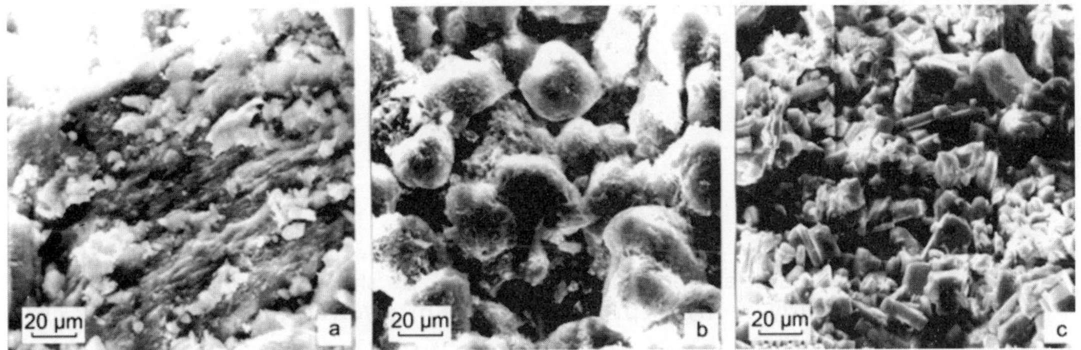

Fig. 10.11(a)   SEM micrographs of Fe/MgO catalyst from magnesium hydroxide (a), from basic Magnesium hydroxycarbonate (b) and from magnesium oxalate (c) (×480) (reproduced from reference G.A. Lomic, E.E. Kis, G.C. Boskovic and R.P. Marinkovic-Neducin, APTEFF, 35,67-77(2004)).

10.6 PARTICLE SIZE AND NUMBER

In heterogeneous catalysis, frequently used catalysts are metal particles supported on an oxidic material. The size of the metal particles plays a crucial role for the catalyst efficiency and the determination of the size distribution is one of the main tasks of electron microscopy in catalysis. Crystalline metal particles can be visualized by HRTEM. Often, the metal has a high atomic number and is a stronger electron scatterer than the oxide (silica or alumina). As an example, the presence of silver particle supported on ZnO is shown in Fig. 10.12. The size, shape and orientation of the silver particle are clearly discernable in this micrograph.

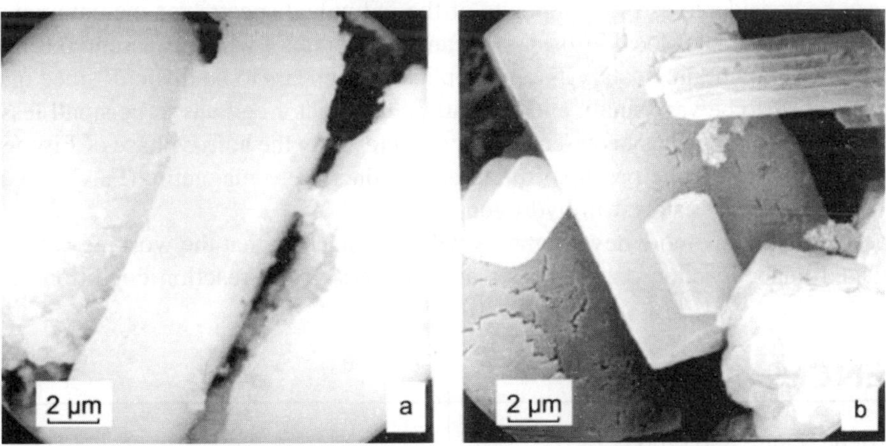

Fig. 10.11(b)   SEM micrographs of the Fe/MgO catalysts, support are originated from magnesium nitrate (a) and from magnesium acetate (b) (×4800) (reproduced from reference G.A. Lomic, E.E. Kis, G.C. Boskovic and R.P. Marinkovic-Neducin, APTEFF, 35 (2004) 67)

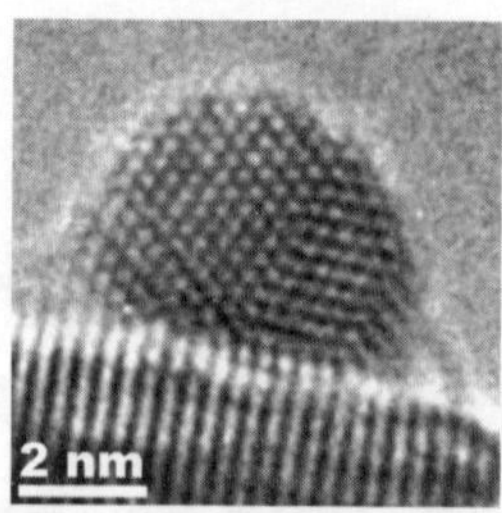

Fig. 10.12 A silver particle supported on ZnO (the size shape and orientation features are clearly seen) (reproduced from http://www.microscopy.ethz.ch/catalysis.htm)

CONCLUDING REMARKS

Many methods have been used to try to achieve high resolution imaging of catalysts. Transmission electron microscopy (TEM) enables the structure of catalyst components to be observed with a spatial resolution hitherto unknown of a fraction of a nanometer. However, it must be remembered that the imaging and elemental compositions have been acquired under vacuum conditions although the actual catalyst faces more severe experimental conditions. It is possible to make *in situ* observations at pressures of up to 10^{-2} bar (1/100 of an atmosphere). However, what is required is the acquisition of TEM images at pressures above 1 bar and above conditions more near to conditions existing in industrial catalytic reactors, still remains extremely challenging. There is various combination tools including spectroscopy to the microscopy have emerged but still most of them still require observations to be made under vacuum.

The recent introduction of Scanning transmission X-ray microscopy (STXM) wherein the sample is scanned with an X-ray beam and the absorption is plotted as a function of X-ray energy. The data generated can be considered as the composite of the individual spectra for the components of the catalyst. Even though precise local information can be generated (which is essential to identify the active centre of the functioning catalysts), the application appears to be difficult since one requires the catalyst particles to be very small, and the path in the reactor cell has to be small less than 100 micrometers. This technique has been recently applied to study the active phase of Fischer Tropsch catalyst wherein the iron oxide precursor undergoes reduction to magnetite (Fe_3O_4) as a result of reduction of the precursor catalyst in hydrogen.

It is expected that many more developments will take place so that the working catalysts can be directly imaged under reaction conditions or at least under semi-reaction conditions in the near future.

REFERENCES

1. A.T. Bell, Nature 456 (2008) 185
2. M. Pan, Micron 27 (1996) 219
3. O. Terasaju, T. Ohsuna, N. Ohnishi and K. Hiraga, Current Opinion in solid state and Materials Science, 2 (1997) 94
4. P.L. Gai and E.D. Boyes, "Electron Microscopy in Heterogeneous Catalysis, Taylor and Francis, (2009)

5. A.K. Datye, J. Catal., 216 (2003) 144

6. M. Jose Yacaman, J.A. Ascencio, S. Tehuaeanero and M. Martin, Topics in Catalysis, 18 (2002) 167

7. J.G. Buglass, A. Howie, and D.W. McComb, Catalysis Letters, 3 (1989) 17

8. J.M. Thomas and O. Terasaki, Topic in Catalysis, 21 (2002) 155

9. J.M. Thomas and P.L. Gai, Advance in Catalysis, 48 (2004) 171

Computer Modelling and Theoretical Aspects of Catalysts

11.1 INTRODUCTION

The use of computational techniques to study the structure, reactivity and technological applications of catalytic materials is now an established branch of modern chemistry [1-3]. There have been many successes in the application of computer modelling techniques to biological systems particularly in solving protein structures which act as enzyme catalysts [4]. The application of theoretical methods to polymer molecules, inorganic complexes, high temperature superconductors, nanomaterials and other materials is equally strong due to accuracy of these method in determining structure, energetics and dynamical behaviour of these complex materials [3]. Molecular modelling now is a growth area in chemistry due to the demands of chemists to be able to:

- ❖ Visualize molecules.
- ❖ Study the structure of molecules.
- ❖ Study the properties of a molecule.
- ❖ Compare the structure and properties of molecules.
- ❖ Study interactions between molecules.
- ❖ Study reaction mechanisms.
- ❖ Predict the structure of molecules.
- ❖ Predict the properties of molecules.
- ❖ Predict reaction mechanisms.

The elementary processes occurring during catalysis are:

❖ Transport of reactant to the active sites.
❖ Sorption of reactant at an active site.
❖ Reaction at the active site and conversion to products.
❖ Desorption and transport away from the active sites.

Hence, a fundamental understanding of catalysis requires accurate knowledge firstly of the structure at atomic level of the catalyst and secondly of structural changes accompanying the key processes such as activation and deactivation.

In the first step of computer modelling of a catalyst we require atomistic models of the active sites of the catalyst and the ways in which molecules interact at these sites. Once accurate models of these structures have been developed we can proceed to unravel the dynamical aspects involving diffusion and reaction which are at the heart of catalysis.

In this chapter we focus on the main methodologies used in contemporary computer modelling studies on catalytic materials and some applications of these techniques to catalytic materials.

11.2 METHODS

There are two broad classes of method used in current computational chemistry:
   (a) Force field (Classical Mechanics) methods
   (b) Ab initio (Quantum Chemical) methods

(a) Force Field Methods: In the force field approach knowledge of electronic structures is subsumed into effective interatomic potentials (V), which describe in analytical or numerical form the variation in the energy of the molecule as a function of the nuclear coordinates $(r_1.....r_N)$ of N atoms present in the molecule. The interatomic potential for a molecule (system) of N atoms is generally broken into 'pair', 'three-body', 'four-body' and higher order terms:

$$V(r_1........r_N) = \sum_{i,j=1}{}' V^{(2)}(r_i,r_j) + \sum_{i,j,k=1}{}' V^{(3)}(r_i,r_j,r_k) + \sum_{i,j,k,l=1}{}' V^{(4)}(r_i,r_j,r_k,r_l) + ...$$

where the first term refers to a sum over all pairs of atoms, the second term over all triplets and the third over all quartets, with the summation continuing in principle up to 'N-body' terms. The primes indicate that in the summation a given term is only included once (e.g. ij and ji are avoided). V is often approximated by the leading, pair potential term $V^{(2)}(r_i, r_j)$ which depends simply on the distance between the nuclei i and j. The pair potential term is usually decomposed into Coulombic and non-Coulombic terms:

$$V^{(2)}(r_i, r_j) = \frac{q_i q_j}{r_{ij}} + \varphi(r_{ij})$$

with the first term being the Coulomb potential between a pair of atoms with charges q_i, q_j and separation r_{ij}. The non-Coulombic $\varphi(r_{ij})$ are the 'short range' interactions which are attributed to the repulsion between electon charge clouds, van der Waals attraction, bond bending and stretching. The following are commonly used function for two-body interactions:

11.2.1 Bonded Interactions

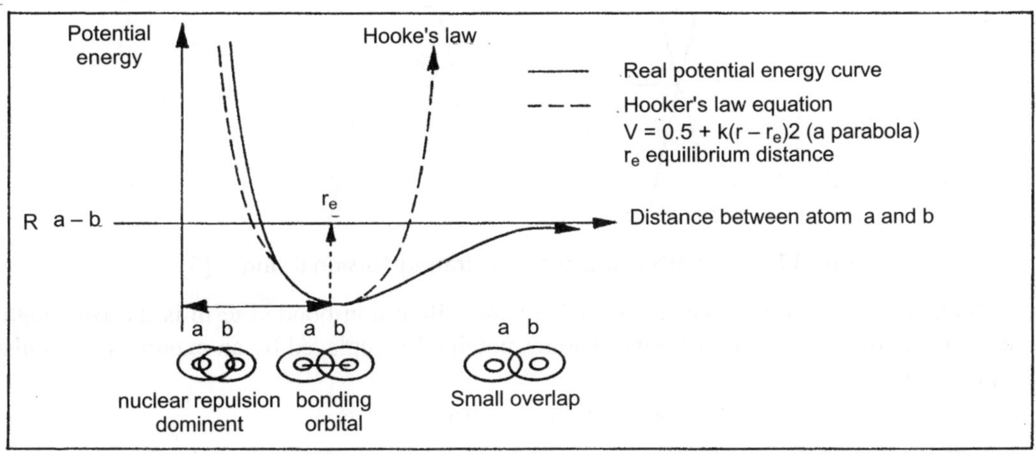

Fig. 11.1 Change in potential energy with respect to bond length

The simplest and most widely used function applied to a bonding pair of atoms is the bond harmonic function, i.e.

$$V = \frac{1}{2}k(r - r_e)^2$$

where r_e is the equilibrium bond distance and k is the bond force constant. Functions of this type are quite adequate for values of r_{ij} close to r_e.

The change in potential energy with bond length is shown in Fig. 11.1. Greater reliability over a wider range of separations can be achieved by use of the Morse function, which has the form

$$\phi(r) = D[1 - \exp\{-\beta(r - r_e)\}]^2$$

where D is the dissociation energy of the bond, r_e is the equilibrium bond length and β is a variable parameter, which can, however, be determined from spectroscopic data.

11.2.2 Non-bonded Interactions

The non-bonded term is usually approximated by analytical functions which generally include both attractive and repulsive components. Several such functions are available, most widely used being the Lennard-Jones potential, which takes the form:

$$\phi(r) = 4\varepsilon\left[\left(\frac{\sigma}{r}\right)^{12} - \left(\frac{\sigma}{r}\right)^{6}\right]$$

with a steeply repulsive r^{-12} function describing the non-bonded repulsion and an attractive r^{-6} term modelling the dispersive interaction. ε is the minimum energy of the function and σ can be interpreted as the approximate radius of the atom.

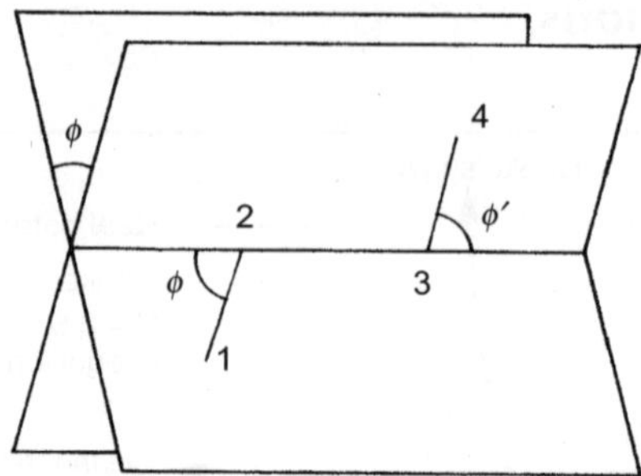

Fig. 11.2 Quartet of atoms illustrating torsional angle [5]

An alternative function that has been used for modelling non-bonded term is the Buckingham potential in which the r^{-12} term of Lennard-Jones potential is replaced by an exponential repulsive term thus giving:

$$\phi(r) = A \exp(-r/P) - Cr^{-6}$$

Lennard-Jones potential have been used widely in modelling rare-gas and molecular crystals. Morse potentials become more appropriate when covalent systems are being studied. Buckingham potentials have been very widely used in the study of ionic and semi-ionic solids.

Inclusion of many-body effects may be achieved by the use of angle-dependent forces. A commonly used example is the simple bond-bending term of the type $V(\theta) = \dfrac{1}{2} K_\beta (\theta - \theta_e)^2$ where θ is the angle subtended by three atoms, θ_e is an equilibrium value and K_β is the appropriate force constant. The use of such terms is most appropriate in systems with directional covalent bonding, such as silicates, where θ_e will correspond to the tetrahedral angle subtended by O-Si-O bonds in the tetrahedral SiO_4 units.

In modelling molecules (especially macromolecules) it is commonly necessary to include a type of 4 body potential that depends on a torsional angle ϕ. Thus in a system of atoms 1, 2, 3 and 4, the torsional angle, ϕ, is that between the planes defined by atoms 1, 2 and 3 and by 2, 3 and 4 as shown diagrammatically in Fig. 11.2

The full potential function is given by:

$$V(\phi) = K_{1234}[1 - \cos(n\phi)]$$

In Fig. 11. 3 the energy profile of ethane molecule with respect to torsion angle is shown.

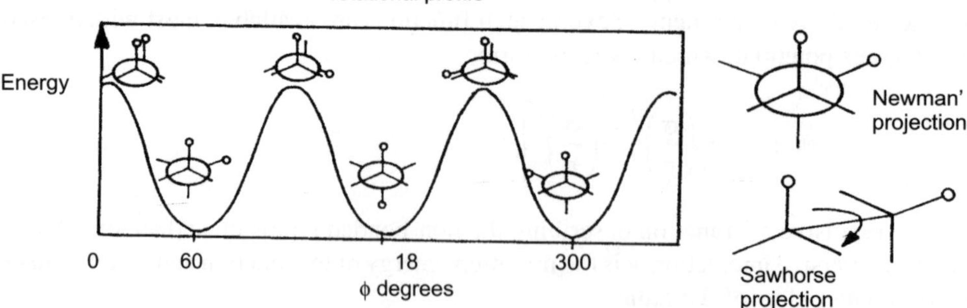

Fig. 11.3 Rotational profile of ethane molecule with respect to torsion angle ϕ

In ionic materials atomic and ionic polarization of solids is also essentially a many-body effect. The simplest approach models polarizability in terms of a point dipole whose magnitude, $\mu = \alpha E$, where α is the atomic (or ionic) polarizability. However, as the dipole is a point entity, it cannot describe the physical basis of polarizability which is the displacement of the valence shell electron density in response to the applied field. Consequently, important physical effects, in particular the coupling of polarizability and short range repulsion, are omitted by these models which fail badly in describing both dielectric and lattice dynamical properties of crystals. These deficiencies are largely overcome by the shell model which describes the development of a dipole moment in terms of the displacement of a massless shell of charge Y (representing the valence shell electrons) from a core (representing the nucleus and core electrons), the core and the shell being connected by a harmonic restoring force for which the string constant is k; the free atom polarizability α is then given by $\alpha = Y^2/k$. Shell model potentials have been very widely used in studies of lattice dynamical and defect properties of ionic solids. [3]

Having defined the form of the interatomic potential, the next step is to determine the variable parameters. There are two principal methods by which potential parameters may be obtained: (i) empirical fitting and (ii) direct calculation. In empirical fitting structural and lattice properties (for solid) are used to determine the potential, by requiring that the potential reproduces these properties. In direct calculation methods the interaction energy between a pair or periodic array of atoms are calculated for a range of distances and the resulting potential curve fitted to a suitable functional form. These methods can, in principle, calculate potentials for any interaction, but may depend on the availability of suitable basis sets.

Having developed a suitable interatomic potential, there is a range of computational techniques which may be employed for calculating structural properties and the dynamical behaviour of molecules, proteins, DNA, polymer, and solid materials.

Force field methods are of broadly classified into three types: *Molecular mechanics, molecular dynamics and Monte Carlo methods.*

Molecular mechanics. Molecular mechanics calculations are a step up from classical energy calculations as they include the energy as calculated from simple potential energy functions considering only non-bonded atoms and also use mathematical techniques [6] to yield a more realistic structure as discussed above.

Bond lengths, bond angles and torsion angles are all altered using potential energy functions during a process known as optimization.

Information obtainable may include

- ❖ Molecular geometry
- ❖ Energy
- ❖ Ionization potential

- ❖ Heat of formation
- ❖ Dipole moment
- ❖ Charge density etc.

This information can be obtained for existing molecules and ones that have yet to be synthesized. It is rare for calculations to include the effects of solvation, aggregation or population distribution. Most calculations are "gas phase" or "in vacuo" calculations.

11.2.2.1 Molecular Dynamics

This method uses the Newtonian equations of motion, a potential energy function and associated force field to follow the displacement of atoms in a molecule over a certain period of time, at a certain temperature and a certain pressure. [5] Calculations of motion are done at discrete and small time intervals and a velocity calculated on each atom position which in turn is used to calculate the acceleration for the next step. Starting velocities can be calculated at random (necessary when starting at 0 Kelvin where the kinetic energy is 0 or by scaling the initial forces on the atoms. Simulations can also be run with given velocities that are chosen with a target temperature in mind. The simulation proceeds by solving the classical (Newton's equation) equation of motion in an iterative manner, in which coordinates (x_i) and velocity (v_i) of the particles in the simulation box are updated after a lapse of a 'time step', Δt, for an infinitesimal value of which we would have the simple update expressions:

$$x_i(t + \Delta t) = x_i(t) + v_i(t)\,\Delta t$$

$$v_i(t + \Delta t) = v_i(t) + \frac{f_i}{m_i}(t).\Delta t$$

where m_i and f_i are respectively the mass and the total force acting upon the ith particle; the forces are, of course, calculated from the derivatives of the interatomic potentials. A finite value in the range $10^{-14} - 10^{-15}$ s is normally used in MD simulations.

When doing calculations on biological molecules it is becoming more frequent to do the calculations in the presence of solvent (usually water!!). However, this brings further complications due to two main problems. The first being increased CPU time due to the larger number of atoms. The second is that the water molecules surrounding the molecule tend to drift away from the molecule of interest and get "lost" from the calculation if only a certain area of space is being monitored as is usually the case. This causes nasty "edge effects". There is one method currently used to get around this problem. That is to place our molecule surrounded in water in a box of a specific size and then to surround that box with an image of itself in all directions. The solute in the box of interest only interacts with its nearest neighbour images. Since each box is an image of the other, then when a molecule leaves a box its image enters from the opposite box and replaces it so that there is conservation of the total number of molecules and atoms in the box. This is known as periodic boundary condition.

Simulated annealing [7] is a special type of dynamics. The molecule is heated and then cooled very slowly so that conformational changes taking place will lead to a local minimum being located. This process is generally repeated many times until several very closely related, low energy, conformations are obtained. These are assumed to be the global minimum.

11.2.2.2 Monte Carlo Techniques

Related to molecular dynamics are **Monte Carlo** methods which randomly move to a new geometry/ configuration but there is no relationship in time between such configurations. Indeed successive configurations are generated by random 'moves' which can be an atomic translation or molecular rotation, or the insertion or deletion of an atom or molecule. In this way, ensembles of configurations are accumulated, from which statistical averages may be calculated. In generating such ensembles, it is essential to formulate an 'acceptance' criterion, i.e. a procedure that determines whether a new configuration created by a move will be accepted within the ensemble. The most commonly used approach is the Metropolis [5] method, which proceeds as follows:

1. Choose the initial position, calculate energy.
2. Pick the random displacement, calculate the new location and energy change, ΔE, associated

with the move. From this the Boltzmann factor B = exp $(-\Delta E/kT)$ is calculated if ΔE is positive. A random number r is generated in the range 0-1.

4. Decide whether to accept the move.
5. If B < r or, $\Delta E \leq 0$, accept the new position, update it and the energy, otherwise, stay at the same place, add the current values to the running averages.
6. Go back to step 2.

(b) Ab Initio Methods: The term *ab initio* means from first principles. These methods are used to compute interactions at the electron and nuclear levels. Traditional Hartree-Fock procedures construct a molecular wave function from individual atomic orbitals and compute the potential energy that arises from the nuclear attraction and the average electronic repulsion from other electrons in their approximate orbitals. The essential idea of the Hartree-Fock or molecular orbital method is that, for a closed shell system, the electrons are assigned two at a time to a set of molecular orbitals. This can be represented by the simple picture:

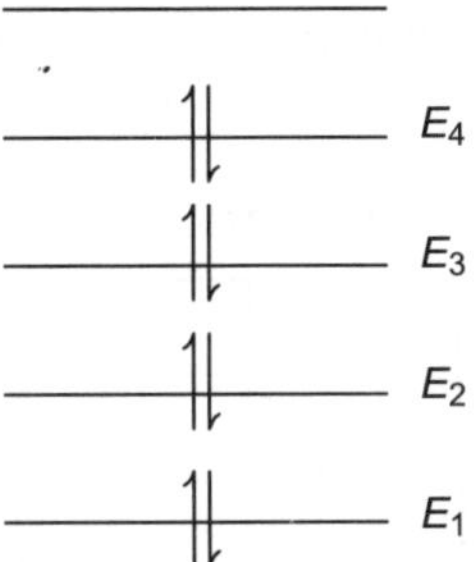

Here we see a system of 8 electrons occupying 4 molecular orbitals of lowest energy. Only one unoccupied molecular orbital is shown. The unoccupied molecular orbitals are often called virtual orbitals.

To give us freedom to vary the molecular orbitals to best suit the molecule in question, we expand each molecular orbital in terms of a set of basis functions which are normally centered on the atoms in the molecule. This gives:

$$\Phi_i = \sum_{\mu=1}^{n} C_{i\mu}\chi_\mu$$

Here each molecular orbital Φ_i is now expanded as a linear combination of basis functions, χ_μ:

Our aim is to find the value of the coefficients $C_{i\mu}$ that gives the best molecular orbitals. The sum is over n basis functions. n is the number of basis functions chosen for the system. We call this 'the basis set size'. The situation for open shell systems (ones with unpaired electrons—often due to an odd number of electrons) is more complex. Post-Hartree-Fock procedures add varying treatments of electron correlation beyond this mean-field approximation.

Density functional theory provides a computationally attractive method for studying catalysis. This approach introduces approximate functionals to describe exchange and correlation. For example, density functionals can be derived from the properties of uniform electron gas. The resulting local density approximation method is particularly effective in treating heterogeneous catalyst systems. Several forms of higher order correction, nonlocal functionals, or generalized gradient approximation methods have been developed which extend the applications of the method. All these quantum chemical methods yield information on chemical structure, energetics, and reactivity of catalytic materials.

11.3 APPLICATIONS

11.3.1 Molecular Mechanics

(a) Conformational Searching: Classical potential energy methods are frequently used when examining many conformations of large molecule. For example, the rotation of one or more bonds in a molecule tries to try and locate low energy conformations. In most biologically oriented modelling this is very important as it is unlikely that high energy conformers are common enough to be involved in biological processes, due to having a Boltzmann distribution of energies.

When two torsion angles are incremented in a stepwise fashion it is possible to plot the energy of each conformer and locate the areas of low energy. This plot is called the Ramachandran plot. The Ramachandran map or Ramachandran plot, indicates low energy conformations for (phi) and (psi), the conventional names for the torsion angles on either side of C's (alpha carbons) in peptides. This plot is often used as a first check on how sensible torsion angles in proteins are. The low energy conformations obtained from Ramachandran plots (Fig. 11.4) are frequently used as starting points for more sophisticated calculations.

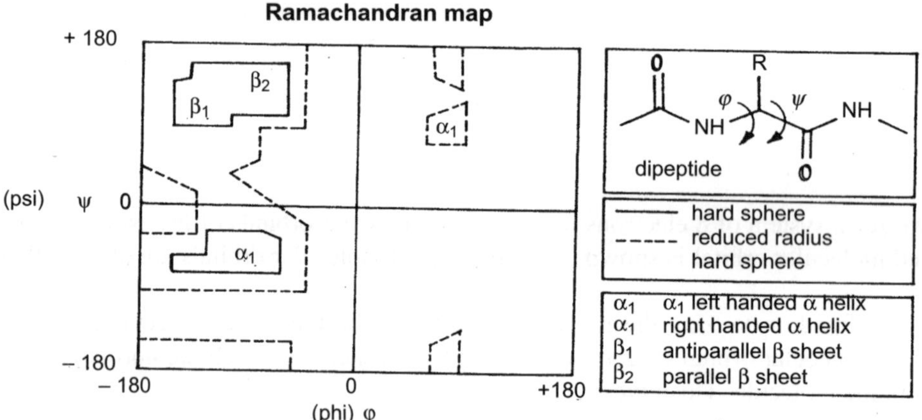

Fig. 11.4 Ramachandran plot for amino acids

(b) Diffusion in Zeolites: Molecular mechanics technique is an efficient method for studying the location and conformation of large guest molecules within the micropores of zeolite catalysts. Here, we demonstrate this approach to study the diffusion behaviour of alkaylbenzenes in zeolites. We use this technique to study the favourable location of ethylbenzene (EB), isobutylbenzene (IBB), *o-*, *m-* and *p*-isomers of isobutylethylbenzene (IBEB) and the isomers of diisopropylbisphenyl (DIBP) in several large pore fully siliceous form of zeolites. [8] The atomic positions of the zeolite framework are held fixed and the molecules are allowed to diffuse through the channel and cavities of zeolites as shown in Fig. 11.5.

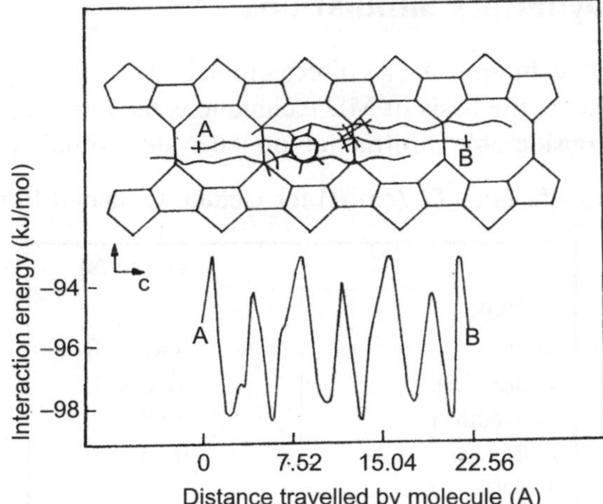

Fig. 11.5 Variation of interaction energy of *p*-IBEB in mordenite zeolite as the molecule diffuses through the 12-m channel. The diffusion tracks for the motion of three carbon atoms are shown

Table 11.1 Diffusion energy barriers in kJ/mol for aromatic molecules in large pore zeolites

Aromatics	Zeolites			
	faujasite	zeolite L	mordenite	ZSM-12
EB	26.92	38.69	6.74	—
IBB	31.65	35.87	10.13	—
m-IBEB	28.38	43.69	17.95	—
o-IBEB	32.74	40.87	95.69	—
p-IBEB	31.65	28.43	6.44	—
4,4′-DIPBP	—	86.66	13.26	34.66
4,3′-DIPBP	—	71.18	24.23	133.62
3,3′-DIPBP	—	79.43	53.31	193.91

The molecule is translated through the 12-m channel of mordenite (MOR) in regular steps of 0.2 Å. After translating the molecule in each step, the most favourable orientation and configuration of the molecule is determined by minimizing the energy of the molecule and interaction energy. Fig. 11.5 includes the diffusion energy profile for *p*-IBEB in zeolite mordenite. The molecule passes through energy maxima and minima when diffusing through the channel from which the diffusion energy barriers are calculated. The diffusion energy barriers of the alkylaromatics are given in Table 11.1 It can be seen that the diffusion energy barriers of the isomers of IBEB are significantly different even though there are small variation in the dimensions of the molecules.

It is seen from Table 11.1 that ZSM-12 is the most shape-selective for the synthesis of 4,4′-DIPBP which is in agreement with experimental results. The selectivity for *p*-IBEB will be maximum in mordenite.

11.3.2 Molecular Dynamics Simulations

Molecular dynamics (MD) technique allows direct simulation of time dependent processes such as diffusion. As discussed above, the basis of MD technique is the integration of Newton's equations of motion, which yields considerable information on the system simulated.

Table 11.2 Diffusion Coefficients, D, (cm²/s) for Octane Obtained from the plots of Fig. 11.6

	D (cm²/s)
T=300K	
total	4.67×10^{-6}
x-direction	1.98×10^{-6}
y-direction	1.95×10^{-6}
z-direction	0.73×10^{-6}
T=450K	
total	13.87×10^{-6}
x-direction	2.53×10^{-6}
y-direction	10.23×10^{-6}
z-direction	1.09×10^{-6}

The Mean Square Displacement (MSD) simulated at 300 K and 450 K for octane are shown in Fig. 11.6 and the diffusion coefficients calculated from the graph are given in Table 11.2.[9]

It is seen from Table 11.2 that at 300 K there is significant diffusivity of the octane molecules through the sinusoidal channel. However, at 450 K the sorbate molecule diffuses through the straight channel, confirming that the equilibrium between octane molecule diffusing in each channel is temperature dependent.

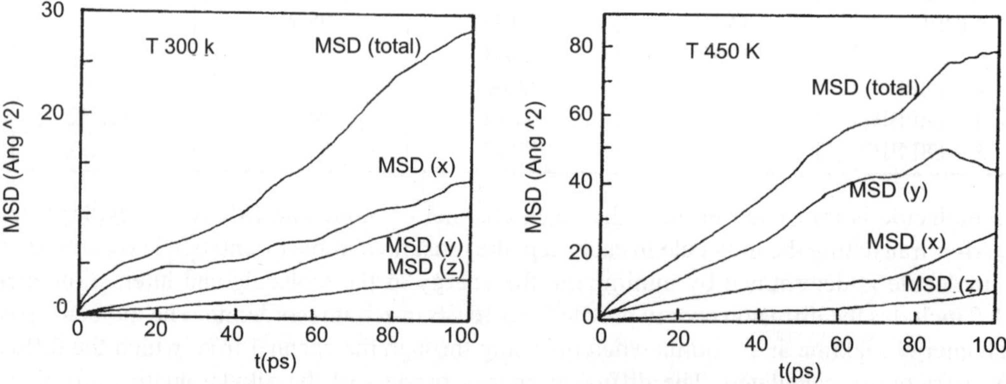

Fig. 11.6 MSD plots for octane obtained from the 100 ps runs at 300K (top) and 450 K (bottom). The overall plots as well as the contributions from each axis direction, x (direction of sinusoidal channel), y (direction of straight channel) and z- directions are shown. The diffusion coefficients obtained from these plots are given in Table 11.2

11.3.3 Simulated Annealing

Amorphous pore walls have been modeled by taking a dense silica glass structure, obtained by simulated annealing, cleaving out a thin layer, approximately three T atoms in thickness and satisfying all dangling Si-O bonds with hydroxy groups.[10] Part of an energy minimised structure is shown in Fig. 11.7 where a Ti atom has been substituted into the pore wall. In this case the Ti-O bond lengths varied from 1.77 to 1.92 Å.

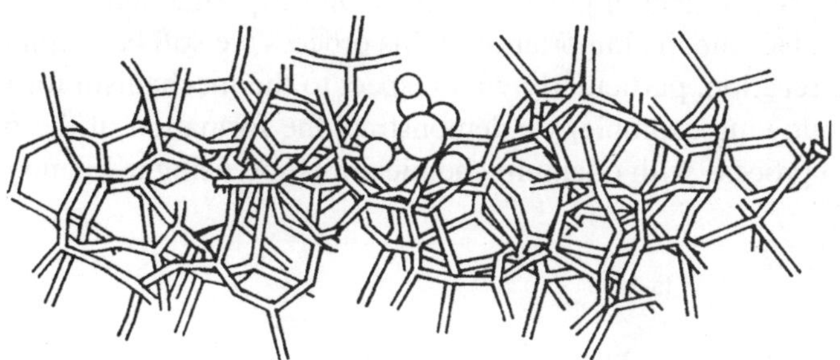

Fig. 11.7 Energy minimised amorphous silica surface, showing the substituted TiO_3OH group

11.3.4 Applying *ab initio* Techniques

Heterogeneous catalysis has been a focus for simulation effort for a relatively long time, with numerous studies of zeolites, metal oxide, and metal catalysts available in literature. But this is an area in which the possibilities offered by new quantum mechanics techniques are particularly exciting, allowing users to study larger, metal-containing systems at a previously impossible accuracy and to simulate the chemical reactions being catalyzed.

There are considerable number quantum chemical program such as GAMESS[11], GAUSSIAN[12], DMol[13], CASTEP[14] etc. where structures, properties, and mode of catalytic action could be studied using these programs.

(a) CO Adsorption:

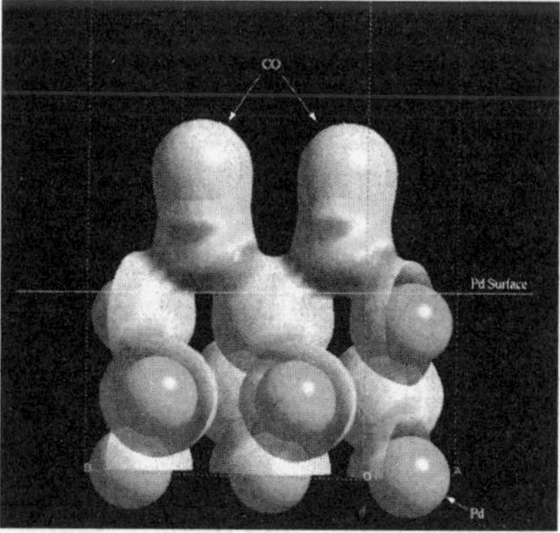

Fig. 11.8 Chemisorption of CO molecules on a typical catalytic surface – as simulated by the CASTEP program. The charge density difference is shown mapped onto the charge density isosurface. This shows valence charge redistribution relative to the superposition of free atoms

An example is the successful modelling of the chemisorption of carbon monoxide on the surface of a typical palladium catalyst (Fig. 11.8)[15]. The CASTEP-computed chemisorption energy agrees closely with the experimentally obtained value. Calculation and display of properties like charge density and electrostatic potential allowed researchers to understand the likely mechanism of attack during any reaction with the chemisorbed molecules.

(b) Methanol to Gasoline (MTG) Reaction: The methanol to gasoline process[16] in medium pore zeolites has been the subject of great theoretical and experimental interest due to its potential industrial importance. Despite the importance of this process we still have limited knowledge of the precise nature of the reaction, particularly with respect to the mechanism for formation of the first C–C bond. In the following example, we demonstrate the formation of the first C–C bond in the methanol to gasoline process with quantum chemical method using a 1T model, $(HO)_3Al(OH_2)$ or ZOH of zeolite.[10]

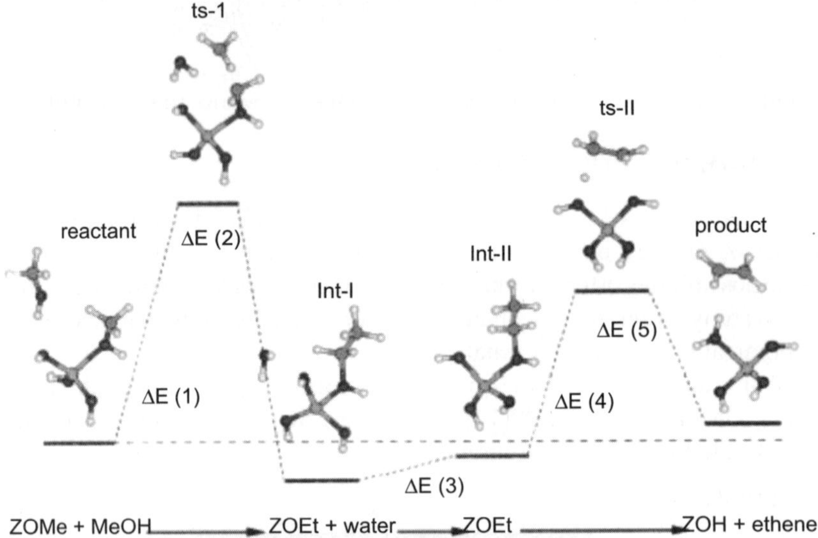

Fig. 11.9 HF/6-31G** reaction parth for formation of ethene from a surface methoxy (ZOMe) and adsorbed methanol (MeOH) via a carbene-like transition state (ts-I) and surface ethoxy intermediates (int-I and int-II)

The formation of methylated analogue of 1T model, $(HO)_3AlO(Me)H$ or ZOMe, ethylated surface, $(HO)_3AlOEtH$ or ZOEt and ethene are shown in Fig. 11.9. The reaction path shown in Fig. 11.9 describes the deprotonation of a surface methoxy group leading to C–C bond formation (in ZOEt) via a transition state with a surface stabilized carbene like species (ts-I). Fracture of the C–H bond occurs early in the reaction coordinate to form water with the OH of methanol and this bond fracture is probably the main contributor to the activation barrier. Deprotonation of this ZOEt species results in the formation of adsorbed ethene through the transition state ts-II.

11.4 SUMMARY

Several computational techniques now being pioneered will enrich the catalyst simulation field. Interesting developments include improved interatomic potentials for organic-inorganic interactions, techniques for computing the diffusivities of slowly moving molecules, and new routes to structure solutions from analytical diffraction data. It is also clear that quantum chemical methods have benefited catalyst studies significantly in understanding the catalytic mechanism. Currently, molecular modeling methods are used for virtual screening before more costly experimental activities are pursued.

The four main families of molecular modelling methods (energy minimization, Monte Carlo, molecular dynamics, and quantum mechanics) all have value in the discovery and development of homogeneous and heterogeneous catalysts. There are significant opportunities for developing and integrating novel and innovative simulation methods into a catalyst system.

REFERENCES

1. *Computer Modelling of Fluids, Polymers and Solids*, (eds. C.R.A. Catlow, S.C. Parker and M.P. Allen), Kluwer Academic Publishers, 1990.
2. *Modelling of Structure and Reactivity of Zeolites*, (ed. C.R.A. Catlow), Academic Press, 1992.
3. *Computer Modelling in Inorganic Crystallography*, (ed. C.R.A. Catlow), Academic Press, 1997.
4. A.T. Brunger and L.M. Rice, In *Adaptation of Simulated Annealing to Chemical Optimisation Problems* (ed. J.H. Kalivas) p. 259, Elsevier, 1995.
5. M.P. Allen and D.J. Tildesley, *Computer Simulation of Liquids*, Oxford University Press, 1987.
6. U. Burkert and N.A. Allinger, *Molecular Mechanics*, ACS Monograph, 1982.
7. S. Kirkpatrick, C.D. Gelatt Jr. and M.P. Vecchi, Science, 220 (1983) 671.
8. R.C. Deka and R. Vetrivel, J. Catalysis, 174 (1998) 88.
9. N. Raj, G. Sastre and C.R.A. Catlow, J. Phys. Chem. B, 103 (1999) 11007.
10. C.R.A. Catlow, L. Ackermann, R.G. Bell, F. Cora, D.H. Gay, M.A. Nygren, J.C. Pereira, G. Sastre, B. Slater and P.E. Sinclair, Faraday Discuss, 106 (1997) 1.
11. GAMESS, General Atomic and Molecular Electronic Structure System, Department of Chemistry, North Dakota State University and Ames Laboratory, Iowa State University, 1999: M.W. Schmidt, K.K. Baldridge, J.A. Boatz, S.T. Elbert, M.S. Gordon, J.H. Jesnsen, S. Koseki, N. matsunga, K.A. Nguyen, S.J. Su, T.L. Windus, M. Dupuis, and *J. A. Montgomery, J.Comput. Chem.*1993, 14, 1347.
12. Gaussian, Inc., Wallingford CT, 2004.
13. DMol, Accelrys, San Diego, CA.
14. M.C. Payne, M.P. Teter, D.C. Allan, T.A. Arias and J.D. Joannopolous, *Rev. Mod. Phys.*, 64 (1992) 1045.
15. P. Hu, D.A. King, S. Crampin, M.H. Lee, M.C. Payne, *Chem. Phys. Lett.*, 230 (1993) 501.
16. S.L. Meisel, J.P. McCullogh, C.H. Lechthaler and P.B. Weisz, CHEMTECH, 6 (1976) 86.

Index